The quantum theory of unimolecular reactions

# The quantum theory of unimolecular reactions

## H. O. PRITCHARD

*York University, Toronto, Canada*

The right of the
University of Cambridge
to print and sell
all manner of books
was granted by
Henry VIII in 1534.
The University has printed
and published continuously
since 1584.

## CAMBRIDGE UNIVERSITY PRESS

*Cambridge*

*London   New York   New Rochelle*

*Melbourne   Sydney*

Published by the Press Syndicate of the University of Cambridge
The Pitt Building, Trumpington Street, Cambridge CB2 1RP
32 East 57th Street, New York, NY 10022, USA
296 Beaconsfield Parade, Middle Park, Melbourne 3206, Australia

First published 1984

Printed in Great Britain at the University Press, Cambridge

Library of Congress catalogue card number: 83-7381

*British Library cataloguing in publication data*
Pritchard, H. O.
The quantum theory of unimolecular reactions.
1. Quantum chemistry
I. Title
541.2'8    QD462
ISBN 0 521 25711 5

# CONTENTS

# PREFACE

Since the work described in this monograph has evolved separately from the mainstream of research on the theory of unimolecular reactions during the past 25 years, a short personal history, summarising the evolution of my thoughts, would seem to be appropriate. When I became a research student in 1948, Michael Polanyi had just resigned from the Chair of Physical Chemistry in the University of Manchester, and M. G. Evans was about to take his place. M.G. arrived permanently in Manchester the following Spring, and he began to build one of the most remarkable schools of physical chemistry that has ever existed: his brother, A.G., together with Ernest Warhurst, Michael Swarcz, H. A. Skinner, and P. H. Plesch, were already there, and J. H. Baxendale returned from Leeds with him. In no time at all, the assembly had grown to include H. C. Longuet-Higgins, W. C. E. Higginson, J. S. Rowlinson, K. E. Russell, A. F. Trotman-Dickenson and Norbert Uri, and, for a while, Michael Kasha. All of these people worked within the ambit of M.G.'s wide range of interests, and the fame of the department spread like wildfire: so much so, in fact, that it was at times almost impossible to get on with one's research for describing it to visitors, who not only included physical chemists and scientists of other persuasions, but civic dignitaries as well! With contagious enthusiasm, M.G. initiated a concerted on-slaught on the problem of energy transfer in chemical reactions, with a longer-term goal of understanding the nature of primary photochemical processes. One result of this was that towards the end of 1950, part way through my Ph.D. on the thermochemistry of the mercury alkyls, with Hank Skinner, I was put in a room in the newly completed Lapworth Laboratory with Aubrey Trotman-Dickenson and we were told to study energy transfer. T-D had just returned from E. W. R. Steacie's laboratory in Ottawa, and was well versed in the latest analytical techniques, which had advanced significantly during the preceding decade; looking back on them now, they were primitive and quite clumsy by modern-day standards. At that time, the most promising reaction which might provide a clean, clear-cut and unambiguous example of a unimolecular reaction with pressure fall-off appeared to be the thermal isomerisation of cyclopropane. We knew that, with our newer analytical method, we could extend the pressure range by at least an order of magnitude lower than the earlier works on this reaction by Chambers & Kistiakowsky [34.C], and by Corner & Pease [45.C] respectively; so (among other things) we began to build an apparatus to remeasure the rate of this reaction. By October 1951 the excitement was growing, and M.G. assigned R. G. Sowden, just recently graduated, to speed the progress of this experiment.

This frontal attack on the energy transfer problem soon began to yield results: first, in their classic experiment, Russell & Simons measured the rate of recombination of iodine atoms in the presence of an extensive series of foreign gases [53.R]. We then followed a few weeks later with the convincing demonstration that the cyclopropane isomerisation behaved exactly as one would expect of a true unimolecular reaction as the pressure was reduced [52.P] and, in the succeeding months, went on to reinforce this conclusion by measuring the restoration of the fallen-off rate by as many as possible of the gases that had been used in the iodine atom recombination experiments [53.P2]. Virtually simultaneously [52.K], Kern & Walters demonstrated the fall-off in rate for the unimolecular thermal decomposition of cyclobutane into two ethylene molecules, and the restoration of this rate for a limited selection of added gases.

Unbeknownst to us, N. B. Slater in Leeds had been trying for some time to calculate the shape of the fall-off curve with pressure for the cyclopropane isomerisation and, in May 1952, he visited Manchester to talk with M.G. about his results. The drama of that day has already been documented by Noel Slater in his book [59.S2]: suffice it to say that the magnificent agreement between the experimental and the theoretical results was of great mutual benefit – it led to the ready acceptance of our results and it aroused enormous enthusiasm for Slater's theory of unimolecular reactions. Sadly, M. G. Evans was already struggling with a persistent sore throat which went undiagnosed for too long, and he died of cancer on Christmas Day, 1952, before being able to communicate our paper to the Royal Society. By that time, we had completed a parallel set of experiments with cyclobutane [53.P3], but after M.G.'s death the focus of the department on energy transfer soon diminished, and many members of the group moved away to take up other appointments.

I was never altogether happy with Slater's theory because, although it had developed out of the original ideas of Polanyi & Wigner [28.P], I could not see how it could account for the trends in frequency factor along homologous groups of molecules, such as the mercury alkyls [53.G] or the aromatic ketones [56.C; 56.P]. It seemed to me that the comparison of the fall-off behaviour for isotopic variants of the same reactant should show up the imperfections of the theory, but I did not do the right experiments [56.G; 58.L], and the credit for so doing must go to B. S. Rabinovitch and his coworkers [58.R; 60.S; 63.R; 65.R]. Nor did I find the alternative transition state approaches any more satisfying, despite the fact that my exposure to theoretical reaction kinetics came from people whose names will always be associated with the concept of a transition state: Michael Polanyi, the Evans brothers, and Warhurst. Having been entranced by Polanyi's first year undergraduate course on bonding and molecular structure, I found myself imbued with the desire to construct a theory which paid more attention to the properties of the real states of the molecule. My determination to do this, however, was further stimulated by Ernest Warhurst's vivid reminiscences of the battle over the thermodynamic approach to rate theory, which took place at the 1937 Faraday Discussion in Manchester; the printed record [38.F] does only feeble justice to the heat of that debate between Eyring, Fowler, Guggenheim, Polanyi, Weiss, and others, and gives little inkling of the fervour with which Eyring and Guggenheim defended their extreme positions! Eventually, I concluded that one could not hope to understand the dissociation of a polyatomic molecule without knowing how a diatomic molecule was broken by heat into atoms: I began to learn about the properties of Morse oscillators [58.G] and, after spending a year (1957–58) in Norman Davidson's laboratory at Caltech, I was able to make my first serious attempt at the description of the dissociation of a diatomic molecule [61.P].

On the experimental front, by the early 1960s, thermal reactions exhibiting fall-off with pressure were no longer a rarity, but I could not see at that time how further study of the shapes of fall-off curves would be helpful on the theoretical front; so I started to look for other useful experiments. I thought that if we could react methyl radicals with trichloromethyl radicals at various pressures, we might be able to probe the $k(E)$ function for the elimination of HCl from the vibrationally excited methylchloroform intermediates which would be formed, i.e.

$$CH_3 + CCl_3 \rightarrow [CH_3.CCl_3]^* \rightarrow CH_2CCl_2 + HCl$$

Phillip Galvin and I spent several months during 1961 trying to find this reaction, but our analytical capabilities were inadequate [63.G]. A similar reaction was found by my brother, G.O., in the early part of 1962 while investigating radical recombination reactions [64.P], and now many examples are known, including possibly some more-general cases where the ejected molecule could be, for instance, hydrogen cyanide [66.T2]. At about the same time, Alan Kennedy and I examined the thermal isomerisation of cyclopropane down to very low pressures, and we were able to demonstrate that collisional activation by the vessel walls was more efficient than activation by collisions with other reactant molecules [63.K2]. This was the first concrete evidence that strong collisions did not deactivate the reactive molecular states upon every collision.

I moved from Manchester to York in 1965, and became a member of a new interdisciplinary institute known as the Centre for Research in Experimental Space Science. Under the guidance of its founding Director, R. W. Nicholls, the boundaries of 'space science' were interpreted liberally enough to encompass the theory (and practice) of chemical kinetics; the benefit for me was that I became associated with a succession of first-rate graduate students and post-doctoral fellows whom I would never have expected to meet in a traditional chemistry department because their skills were in mathematics, physics, or aeronautical engineering.

By the end of 1972, a second cornerstone of the transition state approach was beginning to crumble significantly, for it was now quite evident that widely different transition states could be assumed for a given reaction, but the Rice–Ramsperger–Kassel–Marcus (RRKM) procedure would give the same result for the shape of the fall-off curve [72.N; 72.R; 74.F; 79.A1]. This, as is now well known, arises through the adjustment of the model after the transition state has been chosen so as to force it to be consistent with the observed high pressure rate constant [72.R; 80.P1]. Perhaps it should have sounded the knell for the RRKM theory, much as the unsymmetric isotopic replacement experiments did for the Slater theory a decade earlier, but there was no other substitute available.

By about 1972 also, we had unearthed most of the salient features of the diatomic dissociation problem [73.A; 73.K], and I now had some appreciation of the properties of the master-equation approach to the chemical reaction problem. Consequently, I spent a sabbatical term at the Physical Chemistry Laboratory in Oxford in the autumn of 1972 planning the beginnings of the state-to-state treatment of unimolecular reactions described in the following pages. The unimolecular dissociation process can be regarded as a queueing problem with essentially three steps:

(i) the collisional activation of ordinary molecules into the reactive energy range;

(ii) the randomisation process(es), by which these energetic molecules assemble this energy into a form of motion which will lead to reaction; and

(iii) the reaction process itself in which a molecule is transformed irrevocably into fragments.[1]

It has been commonplace throughout the history of unimolecular reaction theory to try to escape from the straitjacket of a quantum description and slip over into a classical treatment of the problem at the earliest possible moment! I take the opposite extreme position, that of trying to preserve the discrete quantum properties rather longer than is physically reasonable, for it is a much easier conceptual problem to envisage (say) ten million distinct states within the space of one wavenumber of energy than it is to picture the Lissajous motion of a collection of nuclei in ten orthogonal dimensions. Such a position is tenable only because of the remarkable power of modern matrix analysis, by which one can draw conclusions about the eigenvalues of a matrix, regardless of its dimensions. In its simplest form, however, the theory expounded below is little more than a restatement of existing theory in these discrete terms: no significant structure is ascribed to the molecular relaxation process (i), and processes (ii) and (iii) are not disentangled, yielding a result essentially equivalent to that presented in 1972 by Forst [72.F2]. At appropriate stages during the treatment, each of the three facets of the queueing problem will be examined within a discrete framework, and the queueing problem itself will then be solved. This book is unique, in fact, because there is no other account (as far as I know) which begins with a description of molecular relaxation and then goes on to treat the unimolecular reaction as a perturbation of that process.[2] The result is a description of the unimolecular reaction process of transparent simplicity, far easier to use and to understand than is conventional theory.

In order to keep the book short, and to minimise its cost, I will not enter into any description of the RRKM theory since it has been dealt

1 The isomerisation problem, which has often been regarded as the simpler one in many model treatments, is slightly more subtle, and will be dealt with in detail later.

2 I am reminded here of a reply I once heard from Aneurin Bevan to a heckler at a speech during the 1945 General Election: 'you think because water comes through the tap that it comes from the tap!'

with very thoroughly in two excellent books [72.R; 73.F] and in numerous reviews during the past decade. Likewise, I only mention in passing the recent alternative formulation by Troe [77.T2], which also recognises the importance of involving the nature of the collisional activation process in treating weak collision reactions; beyond that point his approach and mine appear, superficially, to be markedly different, but they are not really and a start has already been made recently [82.S3] which will help to establish the connections between the two. For similar reasons, to try to maintain readability, I will not reproduce here many of the heavier mathematical proofs since they too are readily available in the primary literature; they are not necessary for a conceptual appreciation of the unimolecular reaction process.

Finally, although I have harboured a longstanding passion to understand unimolecular reactions in my own terms, I could never have come this far without the meticulous help I have received from Andrew Yau and from Raj Vatsya. Occasionally, I led, but more often than not they found the way and this is reflected by the fact that their names precede mine on the many of the joint publications between us; we have, alas, perpetrated a number of misconceptions from time to time, and I now take this opportunity to redress all of those that I have uncovered so far.

Toronto
November 1982

# ACKNOWLEDGEMENTS

The production of this monograph has been greatly facilitated by the award of a York University Senior Research Fellowship and by financial support for computing provided both by York University and by the Natural Sciences and Engineering Research Council of Canada (formerly the National Research Council). Moreover, it would have been very different in appearance were it not for the computational ingenuity of Avygdor Moise in generating all of the diagrams.

# 1

## The observed properties of thermal unimolecular reactions

I begin with a brief survey of the general properties of thermal unimolecular reactions, noting the salient features which any minimal theory should at least illuminate, if not explain.

### 1.1 The Lindemann mechanism

We are interested here in two general types of chemical process taking place in the gas phase, an isomerisation

$$A \rightarrow B$$

or a dissociation

$$A \rightarrow 2B \text{ or } B + C$$

If such a process is a unimolecular reaction, then the general pattern of its behaviour will be describable, to zeroth order, by the celebrated Lindemann mechanism [22.L; 23.C]:

(i) It is assumed that molecules must possess more than a certain amount of energy $E^*$ before they can be considered to be capable of reacting, and that these energetic molecules arise through the normal collision processes which are always present in any gas, i.e.

$$A + A \underset{k_1}{\rightarrow} A + A^*$$

(ii) In the normal course of events, thermal equilibrium is attained because these energetic molecules are also destroyed by collision, i.e.

$$A + A^* \underset{k_2}{\rightarrow} A + A$$

By detailed balancing we know that, at equilibrium, $k_1[A] = k_2[A^*]$. Thus, the ratio $k_2/k_1$ is equal to the ratio of the populations $[A]/[A^*]$ at equilibrium, and so $k_2 \gg k_1$; also, because $[A] \gg [A^*]$, we can ignore collisions between A* and A* in the derivation below.

(iii) The interesting molecules may decompose to products by a

1

spontaneous process

$$A^* \xrightarrow{k_3} X$$

where X stands for the products, either B, 2B, or B + C as noted above. We write the total rate of the reaction as

$$r = -d[A]/dt = +d[X]/dt = k_3[A^*] \qquad (1.1)$$

By invoking the usual steady state hypothesis [77.V], we assume that the rate of formation of $A^*$ by process (i) can be put equal to its rate of destruction by processes (ii) and (iii) combined, i.e.

$$d[A^*]/dt = k_1[A][A] - (k_2[A] + k_3)[A^*] = 0 \qquad (1.2)$$

whence $[A^*] = k_1[A][A]/(k_2[A] + k_3)$, and the rate becomes

$$r = \frac{k_3 k_1[A][A]}{(k_2[A] + k_3)} = \frac{(k_3 k_1/k_2)[A]}{1 + k_3/k_2[A]} \qquad (1.3)$$

As we will see in a moment, such a reaction will always obey a first order kinetic rate law and, therefore, it is convenient to define a first order rate constant for it as

$$k_{uni} = -[A]^{-1}d[A]/dt = \frac{(k_3 k_1/k_2)}{1 + k_3/k_2[A]} = \frac{k_1[A]}{1 + k_2[A]/k_3} \qquad (1.4)$$

This is known as the Lindemann form, and since the units of $k_1$, $k_2$ are [1/(concentration × time)] and those of $k_3$ are [1/time], $k_{uni}$ has units of [1/time], usually $[s]^{-1}$. At very high pressures, when [A] is very large, the first form for $k_{uni}$ in (1.4) simplifies to $(k_3 k_1/k_2)$, i.e. $k_3$ times the equilibrium concentration of $A^*$, and so it is independent of pressure; this limiting value of the high pressure rate constant is usually denoted by the symbol $k_\infty$. At the opposite limit of very low pressure, the denominator in the second form of equation (1.4) goes to unity and $k_{uni}$ becomes $k_1[A]$ which is, of course, directly proportional to the pressure of the reactant A. At these two extremes we say that the reaction is at its first order limit, or at its second order limit respectively; notice, however, that at the second order limit the reaction does not obey the textbook second order rate law for a rather subtle reason. Processes (i) and (ii) are not 'chemical' reactions in the normal sense: for example, we could equally well write process (i) as

$$A + M \xrightarrow{k_1'} A^* + M$$

likewise the reverse with rate constant $k_2'$, and by retracing our steps the Lindemann expression for the rate constant becomes

$$k_{uni} = \frac{k_1'[M]}{1 + k_2'[M]/k_3} \qquad (1.5)$$

Suppose that we start with pure A, and the reaction is an isomerisation: as the reaction proceeds, B is formed as A is consumed, and it always happens that the rate constant $k_1'$ for the formation of A* from A + B is so nearly equal to $k_1$ for the formation of A* from A + A that it has not yet been possible to detect a drift in the value of $k_{uni}$ as the reactant A is used up and replaced by the product B [53.P2]. Remarkably too, in a dissociation reaction such as the formation of two ethylene molecules from cyclobutane, it happens that an ethylene molecule is just about half as efficient as a cyclobutane molecule in processes (i) and (ii), so that although the pressure rises as the reaction proceeds, $k_{uni}$ remains constant to within experimental precision [52.K; 53.P3]. Finally, putting the Lindemann expression in its most convenient form, we have

$$k_{uni} = \frac{k_\infty}{1 + k_\infty/k_1[M]} = \frac{k_\infty}{1 + k_\infty/k_1 p} \qquad (1.6)$$

where, in the first expression, $k_1$ is in concentration units, or in the second, in units corresponding to the pressure $p$; if $k_1$ refers to the rate of accumulation of internal energy in the pure gas, then, strictly, $k_{uni}$ is the initial rate constant, but from our observations above, it can be used with safety at any instant during the whole of the reaction period. Alternatively, if $k_1$ is interpreted as a suitable average for all the species present in the gas mixture at the moment in question, then $k_{uni}$ will always be the true rate constant.

## 1.2 The experimental patterns

Although it is rarely possible to study any particular unimolecular reaction all the way from the first order (high pressure) limit to the second order (low pressure) limit, many genuine unimolecular reactions have now been characterised over at least part of the fall-off region [72.R]. Thus, we can easily compare the observed *shape* of the fall-off curve, and its *position* on the pressure axis with the behaviour suggested by the Lindemann mechanism.

### (i) *The pressure at which the fall-off occurs*

It is convenient for this purpose to plot the data in a reduced form, that of $k_{uni}/k_\infty$ v. pressure; moreover, because of the large ranges of numbers encountered, and because of the ease of recognising the second order regime in this form, these quantities are always displayed on a log–log plot. This is done in Figure 1.1 for a selection of reactions, where the reactant molecule comprises 3, 6, 9, 12 and 15 atoms respectively. We can see immediately that the larger the reacting molecule, the lower is the pressure at which the fall-off is observed. We define a pressure $p_{\frac{1}{2}}$, known as the 'half-pressure', for which reaction proceeds at exactly one-half of its high pressure rate; at this point, $k_1 p_{\frac{1}{2}} = k_\infty$ and $p_{\frac{1}{2}} = k_3/k_2$. So far, we have no information about the magnitudes of $k_2$ or $k_3$, but we can make a rough guess at $k_2$: this is the rate constant for relaxation of the internal energy of the molecule, and for most polyatomic molecules, must correspond to a time scale of a few tens of collisions [77.L1]. Let us make an extreme assumption, that $k_2$ corresponds to the collision rate itself, usually denoted by $Z$ (the collision number);[1] this is what is generally known as the 'strong collision' assumption. Thus, $p_{\frac{1}{2}}$ is, very roughly, of the order of $k_3/Z$. Neglecting the small variations in $Z$, due to differences in molecular size and mass, or to the differences in the experimental temperatures, then inspection of Figure 1.1 reveals immediately that $k_3$ decreases markedly as the molecular complexity increases. Since the collision rate is of the order of $10^7$ Torr$^{-1}$ s$^{-1}$ in all these systems, the apparent lifetimes of the energised molecules represented in this diagram range from about $10^{-12}$ to about

Fig. 1.1. Behaviour of the fall-off in rate of a unimolecular reaction with molecular complexity. The experimental data are as follows: $n = 3$, nitrous oxide at 2000 K [66.O]; $n = 6$, methyl isocyanide at 504 K [62.S]; $n = 9$, cyclopropane at 765 K [53.P2; 54.S]; $n = 12$, cyclobutane at 721 K [53.P3; 54.S] or methyl cyclopropane at 741 K [60.C]; $n = 15$, ethyl cyclopropane at 741 K [65.H].

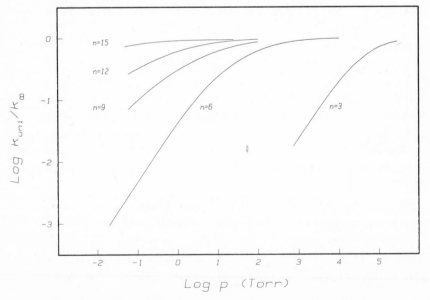

1  Often, $\omega \equiv k_2[M] = Z[M]$, called the collision rate, is used.

$10^{-5}$ s as the number of atoms increases from 3 to 15. If a molecule has sufficient energy to be in the interesting range, the larger it is the harder it appears to be for it to organise its motions into a suitable form so that it will change into another molecule or dissociate into fragments. The basic reason for this is as follows: we require some exactly specified conditions to be met for a state of the molecule to be a reactive one and as the molecule becomes larger, such states constitute a smaller and smaller fraction of the total number of molecular states at the energy of interest; we will see how this happens later on. Such a concept has always been present, either implicitly or explicitly, in all theories of the unimolecular reaction process.

### (ii) Shapes of fall-off curves

Figure 1.1 also reveals the second principal feature of the unimolecular fall-off phenomenon. In the simplest case ($N_2O$), the fall-off curve is rather close to the ideal Lindemann shape; however, as the reacting molecule becomes more complicated, the shape of the fall-off curve becomes flatter, i.e. the range of pressure over which the transition from first order to second order kinetics occurs becomes progressively wider. Without repeating the whole history of the subject, the reason for this flattening effect arises principally,[2] as we will see later, from the fact that all molecules A* do not decompose with a single rate constant, but that the rate increases with increasing energy $E$, so that our $k_3$ has to be replaced by a function $k(E)$, known as the 'specific rate function'. In fact, the *shapes* of the fall-off curves for all of these reactions (except $N_2O$) can be computed moderately well if we simply use the expression

$$k(E) = A_\infty [(E - E_\infty)/E]^{s-1} \qquad (1.7)$$

where $s$ is a number of the order of 1.5 times the number of atoms in the reacting molecule, and $A_\infty$, $E_\infty$ are, respectively, the observed frequency factor and activation energy for the reaction at its high pressure limit. We can assert this from the benefit of 50 years of experience with what is generally termed the classical form of Rice–Ramsperger–Kassel (RRK) theory [32.K; 72.R], and whose premises are usually stated rather more delicately than I have just done.[3]

2 I say 'principally' because there may be other reasons for the flattening that occurs between the 3-atom and the 6-atom case, see Chapter 8.

3 A major stumbling block for RRK theory has been that it is often not possible to find a unique value of the Kassel parameter $s$ which will give both the correct shape of the fall-off curve and the correct position on the pressure axis [72.S2]; the reason for this has been analysed recently [82.S1].

Notice also that different isotopic variants of the same molecule exhibit small but measurable differences in fall-off behaviour; these differences are not easily described within as simple a framework as is represented by equation (1.7).

### (iii) *The addition of inert gases*

It is clear from the discussion of the Lindemann mechanism I have just given that collisions with *any* molecule which happens to be present[4] will assist in the formation and destruction of A*. Consequently, the simple device of adding some inert gaseous substance to a reaction which is taking place at a fallen-off rate will cause that rate to be raised towards its high pressure limiting value. In the early days, this effect was patiently sought after, but analytical difficulties were always prominent; as we now know, it was first found for cyclopropane to which hydrogen had been added [45.C], but the original observation was not so interpreted, for two reasons. The authors themselves thought there were better ways of explaining the changes in rate with pressure in this reaction; the total fall-off in the rate amounted only to about 40%, and the restoration of the rate by the added hydrogen was too small to be convincing. Early experiments were generally inconclusive because those gases which are easily separable from the reaction mixture tend to be rather ineffective in restoring the rate, whereas those which are more effective, given the analytical methods then available, could only be separated with difficulty. With our improved methods, we were able to demonstrate such behaviour for about a dozen unreactive additives in the cyclopropane[5] and cyclobutane reactions [53.P2; 53.P3], and with modern chromatographic techniques, Rabinovitch and his coworkers have measured the relative efficiencies of a remarkable 101 gases in restoring the fallen-off rate of the thermal isomerisation of methyl isocyanide [70.C]. These relative efficiencies are most unambiguously defined on a pressure-for-pressure basis, and this definition is illustrated in Figure 1.2. Suppose that we measure the rate constant $k_a$ for reaction of cyclopropane at a pressure $p_a$. We then add an amount $p$ of inert gas: the new rate constant we actually find, $k_b$, is one which we would have found if we had used the pressure $p_b$ of cyclopropane alone, and the efficiency of the non-reacting added gas is given by $(p_b - p_a)/p$. Alternatively, it is convenient to imagine a relative collision efficiency [53.P2; 70.C], but this is somewhat ill defined since it requires a knowledge of the collision cross-sections of the molecules in

---

4 Or, for that matter, with the wall as we will see later in this chapter.
5 Convincing corroborative evidence comes from an experiment using mixtures of cyclo-propane and monotritiated cyclopropane: there is a small isotope effect, and the two rates exhibit slightly different fall-off characteristics; when gas is added to a reacting mixture at low pressure, not only is the rate raised to some new value, but the isotopic ratio also reverts to that which would have been observed in the undiluted mixture at the particular pressure which would react at that rate [57.W].

question. Finally, it is obvious that the nearer to the low pressure limit it is possible to make these measurements, the greater is the sensitivity of the rate to the pressure of additive; this was another reason for the small magnitude of the effect originally found by Corner & Pease [45.C].

### (iv) *The effect of temperature*

We will divide our discussion of the effect of temperature ($T$) on unimolecular reactions into two parts, one the effect of temperature on reactions in general, and another, the variation of the shape of the fall-off curve with temperature. With few exceptions, the rate constant ($k$) of a chemical reaction increases roughly exponentially with temperature: this is the well-known Arrhenius rate law, $k = Ae^{-E/RT}$, where $A$ and $E$ are commonly called the 'frequency factor' and the 'activation energy' respectively, and $R$ is the gas constant in energy units. In practice, experimental values of $A$ and $E$ are deduced by plotting the observed rate constants in the form of $\ln k$ v. $1/T$ whence the intercept at $1/T = 0$ is $A$, and the slope is $-E/R$. Generally, theories of chemical reaction rates lead us to the conclusion that both $A$ and $E$ will change with temperature, although the magnitudes of these changes are usually expected to be small [72.G1]; consequently, it is necessary to study the rate of a reaction over a very wide range of temperature before one can make a meaningful determination of the curvature of an Arrhenius plot. In fact, for unimolecular reactions, it has yet to be demonstrated that any reaction exhibits a significant deviation from the strict Arrhenius behaviour at its high pressure limit; those few reactions which have been examined over long temperature ranges appear to be strict Arrhenius, to all intents and

Fig. 1.2. The basis of the method for determining the relative efficiencies of gases in restoring the fallen-off rate of a unimolecular reaction.

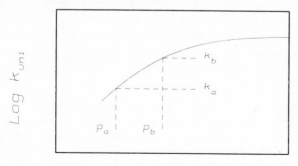

purposes. This observation has important consequences, enabling us to make a good empirical estimate of the $k(E)$ function, as we will see later.

Only a handful of reactions has been observed over long spans of temperature, as follows: The thermal decomposition of di-*tert*-butyl peroxide has been studied over a range of temperature from 363 to 677 K, with a rate constant variation of about ten orders of magnitude, from about $10^{-7}$ to about $10^3$ s$^{-1}$ [68.S; 71.Y; 73.P; 76.L]; likewise, the rate constant for the thermal isomerisation of cycloheptatriene was measured from 600 to 1300 K, with a variation of nearly ten orders of magnitude, from below $10^{-4}$ to about $10^5$ s$^{-1}$ [75.L]. Both of these are very complicated molecules, and we can therefore be certain that the observations were made at the high pressure limit in each case. Almost equally convincing are the results for the thermal dissociation of acetaldehyde [76.E] and for the isomerisation of methylcycloheptatriene [79.H], both of which have been observed over a remarkable 12 orders of magnitude in rate constant; in these cases, however, the high temperature measurements were made somewhat below the high pressure limit and a small theoretical correction was applied to yield the desired values of $k_\infty$, the high pressure rate constant. The isomerisation of cyclopropane is another example: here, the most extended temperature range used in conventional static experiments was from 693 to 808 K at a pressure of 300 Torr, and the results were then corrected to infinite pressure theoretically [61.F1], according to the formula derived from Slater's theory [59.S2]. Shock-tube measurements have been made in the 1300–1500 K range [77.L2], but at pressures which were rather further into the fall-off region; however, a reasonable correction to $k_\infty$ by standard theoretical methods gave results almost coincident with an Arrhenius line drawn through the low temperature data; again, these measurements on cyclopropane span just under 12 orders of magnitude in the rate constant. Finally, somewhat more indirect but nevertheless still reasonably convincing, is the isomerisation of methyl isocyanide: in this case, the rate was measured in the fall-off region (between 2 and 100 Torr) from 393 to 593 K, spanning about seven orders of magnitude in the rate; values of $k_\infty$ were then found by extrapolating the low pressure data according to RRKM theory, and it was shown that the estimated infinite pressure rate data were strict Arrhenius in form to within the probable error of the results [76.C].[6]

---

6 It is now known [80.P2] that these extrapolations underestimated the values of $k_\infty$ by about 30%, but the essential conclusion remains unchanged. Note also that there is a set of shock-tube measurements near 800 K [64.L], but these are so far into the fall-off region that we cannot yet make a correction to infinite pressure with any confidence.

Thus, all the *direct* evidence available leads us to the conclusion that, at the present time, the only supportable position is one of assuming thermal unimolecular reactions to be strict Arrhenius in the high pressure limit. It should be noted, however, that when attempts are made to synthesise a single dissociation rate law from high temperature dissociation measurements, together with low temperature recombination measurements, curved Arrhenius plots invariably result [78.T5; 81.F]; consequently, there is an as yet unresolved conflict here.

We turn now to variations in the nature of the fall-off curve with temperature. Such changes are small [53.P2], and for this reason it is not incongruous to display results for 500 and 2000 K on one graph as we did in Figure 1.1. Instead of comparing the shapes of fall-off curves at various temperatures, which could easily be a rather subjective exercise, the practical approach has been to measure the 'activation energy' at a series of different pressures.[7] Comprehensive information exists for only one reaction, the thermal isomerisation of methyl isocyanide [62.S; 76.C], and in this case the Arrhenius temperature coefficient of the rate falls monotonically from 38.4 kcal mol$^{-1}$ at the highest pressures to about 36.3 kcal mol$^{-1}$ near the low pressure limit (see Figure 5.9; also [78.Y2] for a detailed comparison of both sets of data). Fragmentary results, some more fragmentary than others, exist for several other cases, ethyl isocyanide [69.M1], cyclopropane [60.S; 61.F1; 63.K2; 78.T4; 82.F2], cyclobutane [63.C; 63.V], hydrazine [69.M4], nitrous oxide [66.O], and carbon dioxide [74.W1], but the pattern is always the same, that the activation energy falls as the pressure is reduced. This behaviour, *per se*, does not imply any change in the shape of the fall-off curve with temperature: the main effect is a *very* gradual shift, towards higher pressures, in the position of the fall-off as the temperature is increased. Nevertheless, there are small changes in the actual shapes of fall-off curves with temperature; experimental evidence here is rather sparse, and so I will not belabour the point, but most of our understanding of it comes from numerical experiments by Rabinovitch and his coworkers [65.P; 72.S2]. As we will see in a later chapter, the general pattern is for the fall-off curves to become a little broader as the temperature is raised.

---

7 Quotation marks are used here to emphasise the fact that although the Arrhenius temperature coefficient of the rate, i.e. $E$, has the dimensions of an energy, it is only possible to assign a physical meaning to that 'energy' in certain limiting cases, see Section 5.7.

### (v) *Extremes of pressure*

If we continue to make studies on a particular reaction at successively lower and lower pressures in the same reaction vessel, we will eventually reach a pressure at which the mean free path of the molecules exceeds the linear dimensions of the vessel. Under these conditions, molecules still experience collisions with the walls, but collisions between pairs of molecules become rare events. Energetically excited molecules are then only formed (and destroyed) in collisions with the wall, i.e.

$$A + \text{wall} \xrightarrow{k_1''} A^* + \text{wall}$$

and its reverse. Clearly, $k_1''/k_2'' = [A^*]/[A] = k_1/k_2$. Let $S$ be the surface area of the vessel, and $V$ be its volume: we then have that in the whole volume

$$V\mathrm{d}[A^*]/\mathrm{d}t = k_1''[A]S + k_1[A][A]V - k_2''[A^*]S - k_2[A^*][A]V - k_3[A^*]V$$
$$= 0,$$

whence

$$k_{\text{uni}} = \frac{k_3[A^*]}{[A]} = \frac{k_3(k_1''S + k_1[A]V)}{k_2''S + k_2[A]V + k_3V} \tag{1.8}$$

It follows that $k_\infty = k_3 k_1/k_2$ as usual, and that for very low pressures, $k_{\text{uni}} \to k_\infty/(1 + k_3 V/k_2''S)$; thus, as $S/V$ increases, the limiting low pressure rate tends towards $k_\infty$. The expected behaviour of $k_{\text{uni}}$ (for cyclopropane) is shown in Figure 1.3. The dotted line is the standard fall-off curve, calculated for a hypothetical wall-less reactor, and the solid curve shows how it becomes modified when the reaction takes place in a 1 l sphere; the dashed line is for a smaller reactor with an $S/V$ ratio four times that of the 1 l vessel. Notice that for this example, the limiting low pressure rate increases by only a factor of 2.8 for a four-fold increase in surface to volume ratio. Since $k_2$ and $k_2''$ are simply the gas–gas and gas–wall collision rates multiplied by their respective efficiency factors for de-activation of $A^*$, we therefore have a method for comparing the relative efficiencies of these two kinds of collision. There is only one reaction for which curves such as those shown in Figure 1.3 have been found, and that is the isomerisation of cyclopropane [63.K2], from which it was concluded that gas–wall collisions were three to five times as efficient as gas–gas collisions in the activation–deactivation process; we will examine these data more thoroughly in a later chapter. The beginnings of a turn-up in the fall-off curve have since been reported in several other reaction systems, including ethyl isocyanide [69.M1], cyclobutane [63.B; 63.V;

69.T], and methyl cyclobutane [69.T], with varying degrees of confidence.[8]

Suppose, on the other hand, that we have followed some unimolecular reaction up to its high pressure limit, and we then continue to increase the pressure still further: what might we expect to happen? Most very high pressure experiments in chemical kinetics have been conducted in the liquid phase [41.G; 70.K; 81.I] and the results appear to be interpretable quite satisfactorily in terms of transition state theory: we write that

$$k = e^{-\Delta G \ddagger /RT}$$
$$= e^{(-\Delta H \ddagger /RT + \Delta S \ddagger /R - p \Delta V \ddagger /RT)}$$
$$= A \times e^{-E/RT} \times e^{-p \Delta V \ddagger /RT} \tag{1.9}$$

Consequently, $\mathrm{d}\ln k/\mathrm{d}p = -\Delta V\ddagger/RT$, where $\Delta V\ddagger$ is called the volume of activation, with units (typically) of ml mol$^{-1}$. Generally, unimolecular reactions *in solution* proceed more slowly as the pressure is increased, but

Fig. 1.3. The fall-off in rate of the isomerisation of cyclopropane in an infinite volume (dotted line), in a 1 l sphere (solid line), and in a 16 ml sphere (dashed line), at 765 K.

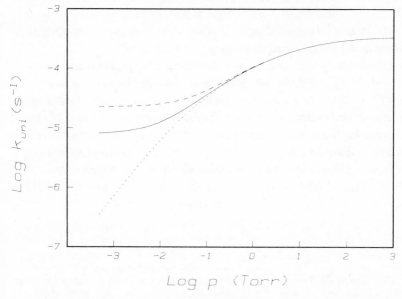

the effect is always minute: if the rate decreases by 10% when the pressure is increased by 100 atmospheres, then that is a large change. In the gas phase, the evidence is both sparse and conflicting. The first example is that of the thermal decomposition of ethylcyclobutane [67.W; 69.A]: in this experiment, the rate constant at 410°C falls by about 10% as the pressure is increased to 170 atmospheres, corresponding to a $\Delta V\ddagger$ of roughly 30 ml mol$^{-1}$; this was thought to be a plausible value on theoretical grounds. Likewise, the thermal isomerisation of cyclopropane has been studied at 482°C up to a pressure of 137 atmospheres [72.J]: the decline in rate here is much smaller, roughly 2% per 100 atmospheres, corresponding to $\Delta V\ddagger = 5$ ml mol$^{-1}$; this is rather close to the change in molar volume between the reactant and the product, which again seems very plausible. Also in the early 1970s, Bunker and his students [72.D4] studied two reactions, those of nitryl chloride and of nitrogen pentoxide, in the presence of up to 300 atmospheres of added nitrogen: the former reaction did not achieve its high pressure limit under these conditions, but in the second one, which was thought to be at this limit, there was a slight indication of an *increase* in rate with increasing pressure.

Finally, we should also note an experiment by Rabinovitch and his students [71.O] in which the thermal decomposition of the *sec*-butyl radical (formed in a chemical activation process) showed no variation in rate from 0.01 to 200 atmospheres of added hydrogen.[9]

We can probably discount the nitrogen pentoxide result on account of the long-standing difficulties in obtaining an accurate measurement of $k_\infty$ itself [81.V4]; thus, we are left with one case in which no effect was seen, but also with two examples of a definite decrease in the high pressure rate with increasing pressure. I have set myself the task of describing unimolecular reaction properties without resort to the concept of a transition state, and although I outline a possible molecular–dynamic explanation for these high pressure phenomena later, the hypothesis still remains to be tested by detailed calculations.

9 For completeness, mention might be made of two other studies, those of di-*tert*-butyl peroxide [68.S] and of methyl isocyanide [70.Y] in the presence of carbon dioxide, although they were not carried out with this purpose in mind; in neither case was any change in rate detected, but the pressure ranges used were too small for us to conclude that there is no effect. This peroxide reaction has also been studied in solution [59.W; 79.B3], and the rate declines slightly with increasing pressure; the effects can be rationalised by assuming values of $\Delta V\ddagger$ of 5–15 ml mol$^{-1}$, according to solvent.

# 2

## The master equation for internal relaxation in molecules

The use of the master equation to describe the relaxation of internal energy in molecules is, in fact, nothing more than the writing of a set of kinetic rate equations, one equation for each individual rotation–vibration state of the molecule.[1] The simplest case we can consider is that of an assembly of diatomic molecules highly diluted in a monatomic gas; under these conditions, we only need to consider the set of processes

$$M + H_2(v_i, J_i) \rightarrow M + H_2(v_f, J_f) \tag{2.1}$$

where the subscripts $i$ and $f$ refer to the initial and final states of the molecule. I have chosen to be specific and to write $H_2$ in this equation because it is now possible to calculate from first principles the collision cross-sections for such processes in this case, at least for low $v$ and $J$ [76.A; 79.A2; 81.M2]. We order the molecular states in some convenient way – for example by energy $\varepsilon$ relative to the lowest accessible state, $v = 0$, $J = 0$ – and denote the initial states with a running index $i$, the final states with the index $j$; for diagrams showing the patterns of energy levels in $H_2$ see [73.K; 77.A; 78.T1]. The rates of formation and destruction of any particular state are then given by the set of simultaneous differential equations for the populations $n_i$, etc.,

$$dn_i/dt = \sum_j (q_{ji} n_j - q_{ij} n_i) \tag{2.2}$$

where $q_{ij}$ is the rate constant for the transition *from* state $i$ *to* state $j$; the first term in the summation takes account of transitions *into* state $i$ from all other states, and the second describes transitions *from* state $i$ to all other states. Notice that it will be convenient to choose our $n_i$ such that $\Sigma_i n_i = 1$, and that the concentration of M has been taken into the $q_{ij}$,

---

1 For the reader who is interested in the more fundamental aspects of the master equation and its application to problems in chemical physics, a valuable collection of reprints has been assembled, together with a lengthy introduction, by Oppenheim, Shuler, and Weiss [77.O]; another good starting point would be the recent book by van Kampen [81.K1].

13

which therefore have units of $s^{-1}$; the numerical magnitude of $q_{ij}$ for any given relaxation experiment is proportional to [M]. By the usual law of matrix multiplication, this equation can be written in a very compact way [58.M]

$$d\mathbf{n}(t)/dt = Q \times \mathbf{n}(t) \qquad (2.3)$$

where $\mathbf{n}(t)$ is a (column) vector of the $n_i$, and $Q$ is a matrix of the rate constants with $[Q]_{ij} = q_{ji}$, $i \neq j$, or (minus) the sum of all the other elements in the $i$th column of $Q$, $i = j$. Thus, the general form for the matrix elements of $Q$ is

$$[Q]_{ij} = [(1 - \delta_{ij})q_{ji} - \delta_{ij}\sum_k (1 - \delta_{ik})q_{ik}] \qquad (2.4)$$

## 2.1 Solution of the master equation

Equation (2.3) describes a set of linear homogeneous simultaneous differential equations with constant coefficients, which is readily solved once the eigenvalues and eigenvectors of $Q$ are known. Let the matrix $Q$ be diagonalised by the transformation

$$C^{-1}QC = J \qquad (2.5)$$

Thus, $C$ is a matrix of eigenvectors, and the diagonal elements of $J$, $j_j = [J]_{jj}$ are the eigenvalues. The solution to equation (2.3) may be written as [55.C; 65.W1]

$$\begin{aligned}\mathbf{n}(t) &= e^{Qt}\mathbf{n}(0) \\ &= e^{CJC^{-1}t}\mathbf{n}(0) \\ &= Ce^{Jt}C^{-1}\mathbf{n}(0)\end{aligned} \qquad (2.6)$$

where $\mathbf{n}(0)$ is the starting population distribution.[2]

We should look at the properties of the matrix $Q$ a little more closely, because there are some further simplifications. To begin with, we know that at thermal equilibrium

$$q_{ij}\tilde{n}_i = q_{ji}\tilde{n}_j \qquad (2.7)$$

where $\tilde{n}_i$, $\tilde{n}_j$ are the equilibrium populations of states $i, j$ at the temperature in question; this is the principle of detailed balancing. Let $E$ be a matrix whose elements are $[e]_{ij} = \tilde{n}_j\delta_{ij}$: then

$$P = E^{-\frac{1}{2}}QE^{\frac{1}{2}} \qquad (2.8)$$

is a similarity transformation which causes $P$ to be symmetric; $P$ has the same eigenvalues as $Q$ and, since it is symmetric, all of them must be real.

---

2 For the reader who is unfamiliar with these concepts, there is a completely worked example for a $2 \times 2$ problem in [70.H1].

In fact, $Q$ is negative semidefinite, having all its eigenvalues negative except one, which is identically zero [58.M; 59.S1]; it is said to be a 'stochastic' matrix. Moreover, in a system like this, there can be no special relationship between the $q_{ij}$ so that all the eigenvalues will be distinct and all the eigenvectors will be unique. The matrix $P$ will be diagonalised by the transformation

$$S^{-1}PS = J \qquad (2.9)$$

and because $P$ is symmetric, $S^{-1} = S^T$. Finally, for reasons which are not apparent now but which will be helpful later, we write

$$R = -P = -E^{-\frac{1}{2}}QE^{\frac{1}{2}} \qquad (2.10)$$

this does nothing except change the signs of all the non-zero eigenvalues from negative to positive, i.e. $\Lambda = -J$. Collecting together the various transformations, we have that

$$S = CE^{-\frac{1}{2}} \quad \text{and} \quad S^T = C^{-1}E^{\frac{1}{2}}$$

whence equation (2.6) eventually becomes

$$\mathbf{n}(t) = E^{\frac{1}{2}}Se^{-\Lambda t}S^T E^{-\frac{1}{2}}\mathbf{n}(0) \qquad (2.11)$$

Discarding the matrix notation now, we may rewrite equation (2.11) in a way which shows how each individual population of the molecule varies with time

$$n_i(t) = \tilde{n}_i^{\frac{1}{2}}\sum_j e^{-\lambda_j t}(\mathbf{S}_j)_i \sum_k (\mathbf{S}_j)_k \tilde{n}_k^{-\frac{1}{2}}n_k(0) \qquad (2.12)$$

where the notation $(\mathbf{S}_j)_k$ means the $k$th element of the $j$th eigenvector of the symmetrised relaxation matrix $R$, corresponding to the eigenvalue $\lambda_j$.

I will use the subscript convention that $\lambda_0$ denotes the eigenvalue which is identically zero, and $\mathbf{S}_0$ is the corresponding eigenvector. If we now examine the long time behaviour of equation (2.12), we find that for sufficiently large $t$, all the $e^{-\lambda_j t}$, $j \neq 0$ drop out leaving

$$n_i(\infty) = \tilde{n}_i^{\frac{1}{2}}(\mathbf{S}_0)_i \sum_k (\mathbf{S}_0)_k \tilde{n}_k^{-\frac{1}{2}}n_k(0) \qquad (2.13)$$

Hence, since the elements of $\mathbf{S}_0$ are given by $(\mathbf{S}_0)_j = \tilde{n}_j^{\frac{1}{2}}$ [58.M], $n_i(\infty)$ simply becomes $\tilde{n}_i$, regardless of the initial population distribution.

## 2.2 The validity of the master equation

The question of whether it is acceptable to use this kind of formulation to describe a relaxation process has been asked on several occasions. Basically, there are two limitations as far as we are concerned here [69.L1; 72.D1]. The first is that we have taken a process which occurs in discrete steps and have written an equation for the evolution of the system as a continuous function of the time. Clearly, if the time $t$ that we are considering is of the order of hundreds of times the mean period between collisions, there will be no significant error.[3] The second is that if the relaxation processes are too fast, the translational Boltzmann distribution will be disturbed; it will then not be possible to define a rate constant $q_{ij}$ corresponding to the temperature of the experiment and, more importantly, the detailed balancing relationship equation (2.7) will not hold. In the case of the rotation–vibration relaxation of $H_2$, relatively small amounts of energy are transferred relatively infrequently, and so there is no difficulty. However, in the unimolecular reaction process, small amounts of energy appear to be transferred at rates comparable with, or faster than, the collision rate [79.M1; 80.P2], and very large amounts of energy appear to be transferred on a time scale of ten or so collisions (see Chapter 5); these are problems which deserve further thought as our understanding of unimolecular reactions improves.

### 2.3 Normal modes of relaxation

It is useful at this point to introduce the compact notation that $( \, , \, )$ denotes the scalar product of two vectors: thus $(S_i, S_j) = 0$, and $(S_i, S_i) = 1$. In this style, equation (2.12) becomes

$$n_i(t) = \sum_j (S_0)_i (S_j)_i e^{-\lambda_j t} (S_j, E^{-\frac{1}{2}} n(0)) \tag{2.14}$$

Equation (2.12) may be separated into a series of independent components, one for each value of $j$. We write

$$n_i(t) = \sum_j (\Omega_j)_i e^{-\lambda_j t} \tag{2.15}$$

where

$$(\Omega_j)_i = (S_0)_i (S_j)_i \times (S_j, E^{-\frac{1}{2}} n(0)) \tag{2.16}$$

If we sum (2.16) over all $j$, we get equation (2.12), for each $i$. It is easy to see that $\Sigma_i(\Omega_j)_i = 0$ for all $j \neq 0$, since $(S_0, S_j) = 0$; this means that each of these independent processes simply shuffles molecules among the energy levels, each with its own rate constant $\lambda_j$ [76.P2], and that the evolution of the population of any individual state is given by

$$n_i(t) = \tilde{n}_i + \sum_{j \neq 0} (\Omega_j)_i e^{-\lambda_j t} \tag{2.17}$$

---

3 One could arrive at the same conclusion by taking a Markov chain approach to the problem of calculating the evolution, see e.g. [79.R3].

The $(\mathbf{\Omega}_j)_i$ of equations (2.16) and (2.17) are dependent on the starting distribution $\mathbf{n}(0)$, and the larger the numerical magnitude of the elements of any $\mathbf{\Omega}_j$, the greater the contribution of the process with rate constant $\lambda_j$ to the overall relaxation. I have written equation (2.16) as a product of two terms $(\mathbf{M}_j)_i \times \zeta_j$, the first of which is the $i$th element of the $j$th normal mode, and the second of which is the flux for that mode, i.e.

$$(\mathbf{M}_j)_i = (\mathbf{S}_0)_i (\mathbf{S}_j)_i \qquad (2.18)$$

$$\zeta_j = \sum_k (\mathbf{S}_j)_k \tilde{n}_k^{-\frac{1}{2}} n_k(0) \qquad (2.19)$$

It is easy to see that if $j=0$, $\zeta_j=1$ regardless of the starting distribution; otherwise, $\zeta_j, j \neq 0$ always has a value which depends on $\mathbf{n}(0)$, unless $\mathbf{n}(0)$ is the equilibrium distribution, when all the $\zeta_j, j \neq 0$ are identically zero.

Figure 2.1 shows the normal modes for the relaxation of para-$H_2$ by helium at 500 K, calculated from the best available theoretical cross-section data [81.M2]. The left-hand panel shows an initial distribution (in this case, for 300 K), and the right-hand panel is the equilibrium distribution at 500 K, $(\mathbf{M}_0)_i = (\mathbf{S}_0)_i (\mathbf{S}_0)_i$; the individual populations are displayed horizontally. The other modes are displayed in order of their rate constants, with the fastest ones to the left; the elements of each mode are normalised to unity, but the overall sign is arbitrary. The fastest mode $\mathbf{M}_9$ corresponds to the transfer of molecules between $J=0$ and $J=2$ of $v=0$ and the next one $\mathbf{M}_8$ is the analogous process for $v=1$. In the relaxation shown here, there is a rather small flux for $\mathbf{M}_8$, given by equation (2.19); physically, this is because this mode does not connect the initial distribution with the final distribution. Then come another pair of processes connecting the $J=0, 2$, and 4 levels of $v=1$, and then of $v=2$, followed by a series of other rotational processes of $v=0$ involving

Fig. 2.1. Normal modes of relaxation for para-$H_2$ at 500 K.

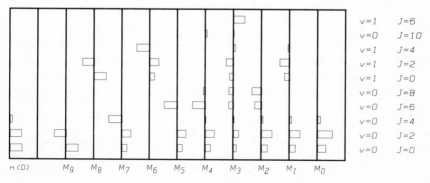

successively higher $J$ values; these become progressively slower because of the increasing difficulty of making transitions between the higher rotational levels [72.P; 73.A]. Finally, we come to the slowest process, $M_1$, which takes molecules simultaneously from $v=1$, $J=0$, 2, 4 to $v=0$, $J=0$, 2, 4; this is clearly the vibrational relaxation.[4] Figure 2.2 shows a similar set of normal modes for para-$H_2$ at 1500 K, again calculated from theoretical cross-section data [78.T1; 79.P2]; the diagram is now much bigger because it is not permissible to truncate the energy-level system at ten levels when describing this relaxation at 1500 K. Nevertheless, the same essential features appear, a progression of rotational processes of increasing relaxation time and, now, two vibrational modes; of these two, the one which couples $v=2$ with $v=1$ is slightly faster than that which couples $v=1$ with $v=0$ because of anharmonicity. A much more extensive discussion of such normal mode diagrams has been given elsewhere [76.P2; 76.P3; 79.P2].

### 2.4 Observed relaxation times

Measurements of relaxation times fall broadly into two classes, those which monitor the populations of some chosen states, and those which measure in some way the impedance of the system to the propagation of a thermal disturbance; many laser experiments fall into the first class, whereas ultrasonic dispersion or shock-tube measurements fall into the second. Although artefacts can occur if unsuitable population v. time profiles are used [76.P3], there is, in general, no real difficulty in using equation (2.14) to obtain the vibrational relaxation rate; we need not discuss this point further at the moment. Problems may well arise, though, in the determination of rotational relaxation rates in this way, as I will show.

Gas–dynamic methods for determining relaxation rates depend upon the measurement of the rate of uptake of energy by the molecules in the path of some thermal disturbance. The shock wave is the easiest example to see, and Figure 2.3 is an attempt to depict the way in which rotational energy is gained by the $H_2$ molecules when the gas mixture is heated from 300 to 500 K, instantaneously. We may write the evolution of the total energy as

---

4 Notice also that in any practical set of normal modes there is a non-physical one ($M_3$ in this case) which arises from the truncation of the levels and the orthogonality requirement on the vectors; there is never, in practice, any significant flux associated with such a mode.

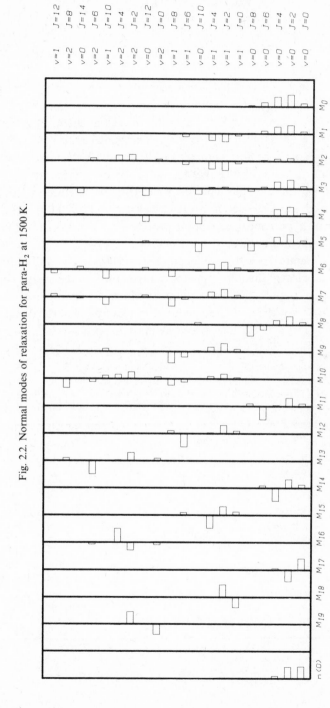

Fig. 2.2. Normal modes of relaxation for para-H$_2$ at 1500 K.

$$E(t) = \sum_i \varepsilon_i n_i(t) = \sum_j \sum_i \varepsilon_i (\mathbf{\Omega}_j)_i e^{-\lambda_j t} = \sum_j \xi_j e^{-\lambda_j t} \qquad (2.20)$$

where

$$\xi_j = \sum_i \varepsilon_i (\mathbf{S}_0)_i (\mathbf{S}_j)_i \times \zeta_j \qquad (2.21)$$

$\zeta_j$ being the flux as calculated in equation (2.18). For $j = 0$, it is trivial to show that $\xi_0 = \sum_i \varepsilon_i \tilde{n}_i$, the equilibrium energy. It is somewhat more cumbersome to show that $\sum_j \xi_j = \sum_i \varepsilon_i n_i(0)$, the initial energy. The vertical axis in Figure 2.3 is the energy (per degree per mole) associated with each normal mode, and the horizontal axis is the characteristic time in seconds (i.e. $1/\lambda_j$) for the mode (notice that the ranges on both axes are very large, and that logarithms are plotted in each case); again, diagrams of this kind are presented and discussed in detail elsewhere [76.P2; 76.P3; 79.P2] for a variety of hypothetical relaxation processes. If we choose a different starting temperature then, from equations (2.18)–(2.21), the resulting energies in Figure 2.3 will be different. This effect is shown in Figure 2.4, in which are plotted the four principal contributions to the energy change for a series of hypothetical relaxations in which the final temperature is always 500 K, but the initial temperature ranges from 300 to 700 K; with a little thought, it is obvious why this must happen given the normal-

Fig. 2.3. Normal mode analysis of the relaxation of para-$H_2$ from 300 to 500 K; the vertical bars depict the logarithm of the energy uptake in units of $R$ ( $= 1.987$ cal deg$^{-1}$ mol$^{-1}$) associated with each normal mode; notice that $\chi_j = \xi_j/\Delta T$, where $\Delta T$ is the temperature change occurring during the relaxation process.

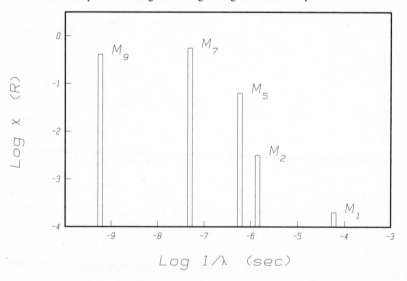

mode diagram (Figure 2.1) and a series of different starting distributions. It thus becomes clear that if one tries to represent the result of such a relaxation measurement in the form

$$E(t) = [E(0) - E(\infty)]e^{-\lambda t} \qquad (2.22)$$

whereas the true behaviour is (say)

$$E(t) = \xi_9 e^{-\lambda_9 t} + \xi_7 e^{-\lambda_7 t} + \xi_5 e^{-\lambda_5 t} \qquad (2.23)$$

the derived value of $\lambda$ will be ill defined and the result of any contrived fit must depend on the starting distribution. Hence, we are forced to the conclusion that rotational relaxation rates will depend upon the method of measurement, whether it is a gas–dynamic one or, because of equation (2.19), a population-following one. The dilemma is much less acute for vibrational relaxation rate measurements for two reasons: first, the various vibrational modes are much more closely spaced in time because of the weaker dependence of cross-section on quantum number; second,

Fig. 2.4. Contributions from the four principal normal modes in the relaxation of para-$H_2$ when the temperature of the gas is changed suddenly from $T_i$ to $T_f = 500$ K. Notice that the total heat capacity declines slightly at the lowest temperatures, as it should for para-$H_2$ [35.F].

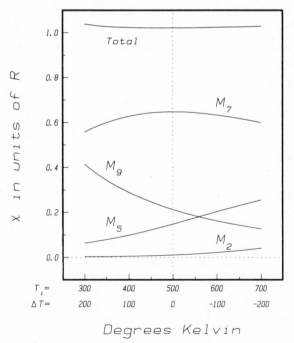

the ground vibrational population is always greater than any other, so that the slowest vibrational mode will always dominate.[5]

## 2.5 Relaxation in non-dilute gases

Although unimolecular reactions are sometimes studied at high dilutions in inert gases, it has been more usual to measure the unimolecular rate for the undiluted reactant. We should therefore investigate, at least briefly, how these simple ideas about molecular relaxation must be modified. Let us extend our discussion of relaxation in hydrogen to a consideration of the undiluted gas: I do this because, again, there is a small amount of theoretical information available for the required cross-sections [74.Z], and so we can still navigate fairly close to the truth. We can now distinguish two types of collision process

[1]      $H_2(v_i, J_i) + H_2(m) \rightarrow H_2(v_f, J_f) + H_2(m)$

[2]      $H_2(v_i, J_i) + H_2(v_i', J_i') \rightarrow H_2(v_f, J_f) + H_2(v_f', J_f')$

In the first of these, the internal quantum state of one of the two molecules remains unchanged, and it is therefore exactly the same as the process we have discussed at length in the preceding pages. In the second, the quantum states of both molecules change simultaneously, which introduces a non-linearity into our set of differential equations; it has not yet proved possible to extend the normal-mode analysis to this case in a satisfactory manner [79.P2], and the conclusions stated below are derived from numerical integrations of the set of rate equations [81.M2]. The simplest thought experiment is to consider a set of relaxations in which $H_2$ is replaced by a structureless molecule M having the same energy transfer characteristics as $H_2(m)$, and let $x$ be the mole fraction of $H_2$ in this hypothetical mixture. Our numerical integrations show that to a very good degree of approximation, the vibrational relaxation rate constant for any such mixture is given by

$$\lambda(x) = \lambda_1 + x\lambda_2 \qquad (2.24)$$

where $\lambda_1$ is the value for infinite dilution of $H_2$ in its structureless replacement, and $\lambda_2$ is the extra contribution to the rate of relaxation from those processes in which there is an exchange of energy between the colliding molecules. The numerical magnitudes of $\lambda_1$ and $\lambda_2$ are about the same: this may seem surprising, since processes of type [2] are usually

5 It is conceivable that at high temperatures, the mean rotational relaxation rate could become comparable with, or even slower than, the mean vibrational relaxation rate, with interesting consequences [76.P3; 79.P2].

thought of as being very fast; however, we are not concerned here directly with the rate at which molecules exchange their internal energy, but with the rate at which internal energy is dissipated as a result of such exchanges. Processes of type [2] are, of course, very fast if they are almost resonant, in which case almost no internal energy is dissipated; on the other hand, they are quite slow if there is a large energy mismatch and the net result is that summed over all the possible contributions, the rate of dissipation of internal energy by type [2] processes is about the same as the average rate for those in which the internal state of one of the pair remains unchanged, i.e. type [1]. Detailed expressions for $\lambda_1$ and $\lambda_2$ in terms of the individual state-to-state rate constants are given in [81.M2]. The important conclusion for us here is that the physical appearance of the relaxation is not altered perceptibly by the inclusion of the exchange processes: we still find a complicated transient behaviour, principally of a rotational character, followed by an exponential decay towards the final equilibrium; this is essentially vibrational in content (just as we saw in Figure 2.1) with the populations of $v=1$, $J=0$, 2, and 4, and of $v=0$, $J=0$, 2, and 4, shrinking simultaneously towards their final values from opposite directions. The only real difference is that the final rate is about twice as fast when the exchange processes are included. I will use these observations later to justify the universal practice of regarding the collision partner in a unimolecular reaction as a structureless entity, even though the reaction being considered is taking place in the undiluted gas.

## 2.6 Relaxation in polyatomic gases

The experimental results for the relaxation of internal energy in polyatomic gases have been well summarised in various places, see e.g. [61.C; 77.L1]. The basic observations are as follows: rotational relaxation rates appear to be commensurate with the collision rate itself and are in general rather poorly characterised, partly because of their great speed, but also because of the intrinsic ambiguity in their definition as I have just shown; vibrational relaxation rates, on the other hand, are quite well defined and reasonably well understood [59.H; 77.L1], and take place on time scales of from a few tens to a few thousands of collisions.

Let us concentrate for the time being on those molecules with fast vibrational relaxation rates: these are generally the so-called strong colliders. If a collection of such molecules is subjected to some perturbation, then their rotational and vibrational populations will return to their equilibrium values through a set of processes whose rate constants may all be within an order of magnitude of each other. If the temperature is high enough for reaction to be a possibility, then there will be many millions of energy levels involved and, at the very least, many thousands of participating normal-mode processes. It is clear that under such conditions, all one could hope to discern would be a single relaxation whose rate constant would be some suitable mean for all those processes contributing to the overall relaxation. The perturbation that interests us is, of course, the reaction and so if we can construct a set of transition probabilities which gives rise to an exponential relaxation, it will be adequate for the description of strong collision unimolecular reaction processes.

For molecules which are regarded as weak colliders, e.g. $N_2O$ or $CO_2$, the rotational and vibrational components of the relaxation process are reasonably well spread out (rather as in our $H_2$ example, except that there are four vibrational degrees of freedom instead of one). We might not expect, therefore, to achieve a good description of the reactions of these molecules by assuming that all the relaxation modes can be lumped into one and, in fact, a proper understanding of such reactions will require a good knowledge of the rotational and vibrational patterns within the relaxation itself. Since at the present time we can only understand fully the internal relaxation of a molecule which is highly diluted in an inert gas, experiments in weak collision systems will be most useful when the reaction rates are measured at high dilutions; conversely, if the range of relaxation eigenvalues is very small, there will be little hope of detecting effects which stem from the *details* of the internal relaxation, and it will not matter much whether these reactions are studied diluted, or undiluted. It so happens, mainly for practical reasons, that weak collision reactions are often studied at high dilution in inert gases, whereas strong collision reactions have usually been observed in un-diluted gases (except when relative collision efficiencies were the object of the measurements).

## 2.7 Pure exponential relaxation

There has been, from time to time, considerable interest in the possibility of pure exponential decay of various observable quantities in a relaxing system; the necessary and sufficient conditions for exponential decay have been formulated in the form of a set of 'sum rules' [63.S2; 64.A], constraints to which the elements of the relaxation matrix must conform, given the energy levels of the system and their degeneracies. If we require only the total internal energy to decay in a pure exponential manner, then the constraints are fairly simple [75.P3], i.e.

$$\sum_j (\varepsilon_j - \varepsilon_i)q_{ji} = \mu[\varepsilon(\infty) - \varepsilon_i] \qquad (2.25)$$

for all $i$; the symbol $\varepsilon$ (without a subscript) denotes the mean energy, per molecule. A matrix of this form has a set of distinct eigenvalues, of which the last two are $-\mu$ and zero respectively. Examination of the properties of such a relaxation [77.Y1; 78.Y1] reveals that although the total energy of the system decays exponentially, the populations do not: the decay of the populations is described, as we have already seen, by a set of normal modes but it so happens that these modes only shuffle the populations without effecting any change in energy; the $\xi_j$ of equation (2.21) are all identically zero except for $j = 1$ and (of necessity) $j = 0$. An even simpler case was discovered independently by Procaccia & Levine [75.P2; 75.P3] and by Yau [77.Y1; 78.Y1] in which both the energy and all the populations decay with a single rate constant: we may write such a set of transition rates in the form

$$[Q]_{ij} = \mu[(1 - \delta_{ij})\tilde{n}_i - \delta_{ij}(1 - \tilde{n}_i)] \qquad (2.26)$$

where $\mu$ is a constant. This matrix has one zero eigenvalue, as it must because it is stochastic, and all the remainder of them are degenerate and equal to $-\mu$; consequently, $\mu$ is the rate constant for the relaxation. Notice that the meaning of equation (2.26) is that the microscopic rate constant $q_{ij}$ for a transition from state $i$ to state $j$ is simply equal to $\mu\tilde{n}_j$, regardless of which initial state $i$ we consider. It follows as a result of this total degeneracy of the non-zero eigenvalues that the relaxation will have some very simple properties [77.Y1; 77.Y3; 78.Y2]

$$\left.\begin{aligned} d\varepsilon(t)/dt &= \mu[\varepsilon(\infty) - \varepsilon(t)] \\ n_i(t) &= \tilde{n}_i + [n_i(0) - \tilde{n}_i]e^{-\mu t} \end{aligned}\right\} \qquad (2.27)$$

and so on. Moreover, the relaxation matrix, equation (2.26), commutes with any other possible relaxation matrix for the same system [79.P3]: because of all these useful properties, I will reserve the name $A$ (with elements $a_{ij}$) to denote the matrix, equation (2.26), whence if we construct a new relaxation matrix

$$[xA + (1 - x)Q] \qquad (2.28)$$

then this new matrix has the same eigenvectors as does $Q$ for $x < 1$, and its eigenvalues are $x\mu + (1 - x)\lambda_j$ where $\lambda_j$ are the eigenvalues of $Q$.[6]

6 It would appear that this would be a useful property when discussing the relaxation of mixtures of gases in which one component is a large polyatomic.

If we substitute equation (2.26) into equation (2.10) to obtain the symmetrised form of $A$, we arrive at a very simple form which may be expressed as

$$R = -E^{-\frac{1}{2}}AE^{\frac{1}{2}} = \mu(1 - p_0) \tag{2.29}$$

where 1 represents the identity matrix and the elements of $p_0$ are given by $[p_0]_{ij} = \tilde{n}_i^{\frac{1}{2}}\tilde{n}_j^{\frac{1}{2}}$; the operator $p_0$ is known as an eigenprojection, formally[7] $S_0(S_0, \ )$, and the ability to recast equation (2.26) in this form provides us with some extremely powerful analytical capabilities.

The transition rate constants described by equations (2.25) and (2.26) may not appear at first sight to be physically reasonable: for example, equation (2.26) must overemphasise the rates of transitions with large changes in energy, or with large changes in angular momentum. On the other hand, I have just argued above that if we could construct a set of transition rates which would give rise to a simple exponential decay, such a set might provide an adequate description for strong collision uni-molecular reactions. This, in fact, turns out to be true as I will show in the next few chapters, where I provide a much more logical development of the material than is available in the literature, essentially because all the pieces of the puzzle did not fall into place in order. The first step was the discovery that a tractable solution for the unimolecular rate could be found if equation (2.26) was used to describe the internal relaxation of the reactant molecules [78.Y1; 78.Y2]. Only later did we realise [79.P3] that equation (2.26) is identical with the strong collision rate constants, as defined by Nordholm [77.N; 78.N], which are actually required in unimolecular theory. Subsequently, Vatsya was able to show [80.S] that Yau's original and rather complicated series solution for the rate [78.Y2] could be put into a closed form which enables us to clarify many of the concepts of unimolecular reaction theory.

It is also possible that the use of equation (2.25) to describe the internal relaxation may have some value for certain unimolecular reaction problems – for example, in cases where it is not admissible to assume exponential decay of the populations in the unperturbed relaxation[8] – but I have not yet had time to examine such a reformulation to see whether any tractable results ensue.

7 The convention here is [column] × [row] × [column].
8 This might be so when it becomes unexceptional to make measurements in the transient or incubation phase of a unimolecular reaction. In general, the transient phase of a relaxation in a closed system may contain overshoots [76.P3] or damped oscillations [81.S2] of certain populations; the number of such oscillations is, however, finite if the eigenvalues of the relaxation matrix are real.

# 3

# *Reaction as a perturbation of the internal relaxation*

Suppose that some of the high energy levels of the relaxing molecule may transform themselves, for example as in a predissociation, into another species: equation (2.2) for the evolution of the populations then becomes

$$\mathrm{d}\eta_i/\mathrm{d}t = \sum_j (q_{ji}\eta_j - q_{ij}\eta_i) - d_i\eta_i \qquad (3.1)$$

where $d_i$ is the rate constant for the spontaneous transformation of molecules in state $i$ into products; we need not concern ourselves with the nature of those products at this stage. I have now written the Greek $\eta_i$ for the populations of the reactant states: thus the amount of reactant remaining at any time $t$ is $N(t) = \Sigma_i\eta_i$, and is only equal to 1 at $t = 0$.

## 3.1 The rate of reaction

The derivation now goes through very much as before. Equation (3.1) is, in matrix form

$$\mathrm{d}\boldsymbol{\eta}(t) = [Q - D]\boldsymbol{\eta}(t) \qquad (3.2)$$

where $D$ is a diagonal matrix of decay rate constants and, of course, the elements of $D$ are all zero below the reaction threshold. The presence of these extra terms on the diagonal does not interfere with the symmetrisation, and we can construct a symmetric reaction rate matrix

$$K = -E^{-\frac{1}{2}}[Q - D]E^{\frac{1}{2}} \qquad (3.3)$$

Consequently, $K$ may be diagonalised by the set of eigenvectors $\boldsymbol{\Psi}$ giving

$$\boldsymbol{\Psi}^{\mathrm{T}}K\boldsymbol{\Psi} = \Gamma \qquad (3.4)$$

the eigenvalues $\gamma_j = [\Gamma]_{jj}$ are all real and positive. By exactly the same set of transformations, then, the evolution of the populations becomes

$$\eta_i(t) = \tilde{n}_i^{\frac{1}{2}}\sum_j \mathrm{e}^{-\gamma_j t}(\boldsymbol{\Psi}_j)_i\sum_k (\boldsymbol{\Psi}_j)_k \tilde{n}_k^{-\frac{1}{2}}\eta_k(0) \qquad (3.5)$$

We will discuss the behaviour of the eigenvalues of the perturbed matrix

$K \equiv -[Q-D]$ shortly: for the present, all we need to know is that (especially at high pressures) they are very little different from those of the unperturbed relaxation matrix $R \equiv -Q$ [77.P; 79.P3]; the most significant change to occur is $\lambda_0$ of $Q$ is identically zero, whereas $\gamma_0$ of $K$ has a small positive value. Thus, at long times, the populations will decay according to

$$\eta_i(t \to \infty) = \tilde{n}_i^{\frac{1}{2}}(\Psi_0)_i \sum_k (\Psi_0)_k \tilde{n}_k^{-\frac{1}{2}} \eta_k(0) e^{-\gamma_0 t}$$

$$= \sigma_i e^{-\gamma_0 t} \tag{3.6}$$

whence the amount of reactant remaining is $N(t) = \Sigma_i \sigma_i e^{-\gamma_0 t}$. The rate constant for a first order chemical reaction is given by

$$k = -[N(t)]^{-1} dN(t)/dt = \gamma_0 \tag{3.7}$$

in other words, $\gamma_0$ is the rate constant.

## 3.2 The topology of the perturbed eigenvalues

Suppose that we calculate the eigenvalues of $R \equiv -Q$ at some convenient pressure: for a weak collider like $CO_2$ at 1 atmosphere pressure, all the eigenvalues might fall within a range, very roughly, of $10^6$ to $10^9 \, \mathrm{s}^{-1}$. By the way $Q$ was constructed, equations (2.2) and (2.3), all the elements are directly proportional to the pressure and so all the eigenvalues of $R$ scale with the pressure; thus, if we make a (log–log) plot of the eigenvalues against pressure, we find a bundle of straight lines of unit slope. Turning now to the eigenvalues of $K \equiv -[Q-D]$, we can see that at high enough pressures, the elements $d_i$ become insignificant and the eigenvalues of $K$ tend to those of $R$; at very low pressures, on the other hand, the elements $d_i$ dominate and they all come through as eigenvalues, with the remainder being small and linear with pressure. Figure 3.1 depicts the actual behaviour of a set of perturbed eigenvalues as a function of pressure. At very low pressures, there is one bundle of eigenvalues, representing the internal relaxations, directly proportional to the pressure as we expect, and another bundle which are the $d_i$. As the pressure is increased, these two bundles approach each other until an 'interaction region' is reached where they merge into a single bundle with (in the limit) unit slope.[1] Notice that the eigenvalue which was zero in the unperturbed matrix, i.e. $\gamma_0$ in the perturbed case, remains very small throughout the whole range of pressures; when the pressure is small, $\gamma_0$

---

1 There is no crossing of any of the eigenvalues in the interaction region: we simply find a manifold of avoided crossings, as in many other eigenvalue problems; further details of the behaviour of these eigenvalues are given elsewhere [77.P; 79.P3; 81.V2].

remains separated from the internal relaxation bundle by a constant factor, but when we reach the pressure where the two bundles of eigenvalues would have intersected, $\gamma_0$ turns over and becomes independent of the pressure.[2] Thus, very crudely, $\gamma_0$ has the correct formal behaviour, and the question we have to ask, and subsequently to understand, is how this interaction between the two bundles of unperturbed eigenvalues (of $Q$ and $D$ respectively) leads to a subtle variation in the shape of the fall-off such as we depicted in Figure 1.1: this we will do in Chapter 5.

### 3.3 The reactant state distribution

We may write the analogue of equation (2.14) as

$$\eta_i(t) = \sum_j (\mathbf{S}_0)_i (\mathbf{\Psi}_j)_i e^{-\gamma_j t} (\mathbf{\Psi}_j, E^{-\frac{1}{2}} \boldsymbol{\eta}(0)) \tag{3.8}$$

which can be partitioned, as before, into a set of flux terms $(\mathbf{\Psi}_j, E^{-\frac{1}{2}} \boldsymbol{\eta}(0))$

Fig. 3.1. Schematic representation of the behaviour of the eigenvalues of the reaction rate matrix as a function of the pressure.

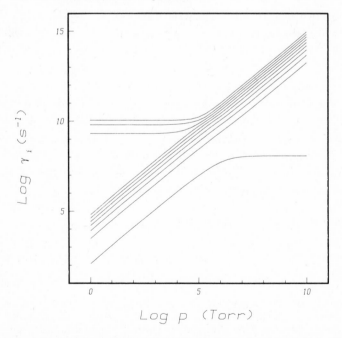

2 The behaviour shown in Figure 3.1 is simply the *n*-relaxation rate generalisation of the 2-relaxation rate problem, see e.g. [72.B1; 77.P].

and mode-like terms $(S_0)_i(\Psi_j)_i$ – see equations (2.18) and (2.19) – together with the appropriate attrition rate term.[3] The properties of equation (3.8) have not yet been examined in any great detail but it is, nevertheless, possible to draw some definitive conclusions about the 'steady' population distribution from which the actual reaction process takes place. Rewriting the long time distribution, equation (3.6), as

$$\eta_i(t) = (S_0)_i(\Psi_0)_i e^{-\gamma_0 t}(\Psi_0, E^{-\frac{1}{2}}\eta(0)) \tag{3.9}$$

we see that we need only to know the perturbed eigenvector $\Psi_0$. Expressions for this are known [78.Y2; 81.P2], the later one being preferable since it can be expressed only in terms of the eigenvectors and eigenvalues of the unperturbed matrix and of the perturbation elements $d_i$; the resulting formula, see equation (8.6), is cumbersome and will not be repeated here. The net result is that $\Psi_0$ is very nearly equal to $S_0$, and differs from it only in some very small subtractive terms which correlate with the $d_i$; consequently, the most significant differences between the two vectors occur in the reactive region, as one would expect intuitively. Thus, the 'steady' distribution which appears to be decaying with the rate constant $\gamma_0$, $\eta_i(0) = (S_0)_i(\Psi_0)_i$, is almost the same as the equilibrium distribution for that temperature, but is diminished to some extent in and near the reactive region. Many examples of such instantaneous distributions are available in the literature [66.T1; 67.R; 68.T; 69.M2; 71.M; 72.G2; 72.R; 75.T; 77.T1; 78.B1; 80.F1], and they will not be reproduced here.

Another conclusion we can draw from equation (3.9) is that the 'steady' reactant distribution retains some memory of the initial distribution, but we should not be misled into reading too much significance into it. In a typical static reaction experiment, where the vibrational relaxation rate constant might be $10^6 \, s^{-1}$ and the rate constant $10^{-4} \, s^{-1}$, the starting distribution can safely be taken to be $\tilde{n}_i$, whence the flux term $(\Psi_0, E^{-\frac{1}{2}}\eta(0))$ will be exceedingly close to unity and the residual fluxes for all other $j \neq 0$ will be infinitesimal. In a shock-tube experiment, on the other hand, there will be some reaction during the transient period, and this will show itself as a series of small values for the $j \neq 0$ flux terms and then, of course, the $j = 0$ flux term appearing in equation (3.9) will be correspond-

---

3 I have previously used the term 'normal modes of reaction' in this context [71.M]: the same term has been used in rather different contexts on other occasions [68.B2; 69.H]; for the time being, at least until their properties have been investigated in more detail, it is perhaps preferable to be more cautious and to refer to the quantities $(S_0)_i(\Psi_j)_i$ as simply perturbed normal modes (of relaxation).

ingly reduced. At the time of writing, it would appear to me that there is scope for a more detailed examination of these matters than I have just given.

### 3.4 Practical evaluation of the unimolecular reaction rate

There must be three basic ingredients of any attempted calculation of the unimolecular reaction rate for a polyatomic molecule. First, we must have a reasonable knowledge of the disposition of the energy levels of the reactant molecule and of their degeneracies, for without this information we cannot know the unperturbed eigenvector $S_0$ whose elements are simply $\tilde{n}_i^{\frac{1}{2}}$ where $\tilde{n}_i$ represents the equilibrium Boltzmann distribution for the temperature of interest. I will take it, without question, that for the purposes of calculating the $\tilde{n}_i$ (and therefore $S_0$), the molecule can be regarded as a rotating collection of independent Morse oscillators without any rotation–vibration interaction. This is obviously a gross oversimplification, but as we will see in the next chapter, there are some strong cancellation effects which shelter us from disaster in this endeavour.

Before we can proceed any further, we must remember that any interesting molecule will possess many thousands, if not many millions of states, and therefore that some kind of graining procedure must be undertaken. For the sake of simplicity, we will imagine the grains to be of equal width in energy $\varepsilon$ (although such an assumption is by no means necessary [78.Y2; 81.P2]), with a running index $r$: thus, the population $\beta_r$ in grain $r$, at equilibrium, is given by

$$\beta_r = \{Q(T)\}^{-1} P_r e^{-\varepsilon_r/kT} \tag{3.10}$$

where $P_r$ is the sum of the degeneracies of all the energy levels within the grain $r$, $\varepsilon_r$ is the energy of the grain, $k$ is Boltzmann's constant, $T$ is the temperature, and $Q(T)$ is the molecular partition function over all of the internal energy states, viz.

$$Q(T) = \sum_r P_r e^{-\varepsilon_r/kT} \tag{3.11}$$

Second, we must know something about the decay rate constants for those of the states which are capable of converting themselves into product. Let us assume that within the $r$th grain the appropriate decay constants are $d_{i,r}$, with corresponding state degeneracies $g_{i,r}$, whence the required mean decay rate constant for the whole grain will be

$$d_r = \sum_i g_{i,r} d_{i,r} / P_r \tag{3.12}$$

this can also be written as

$$d_r = \sum_i \tilde{n}_{i,r} d_{i,r} / \beta_r \qquad (3.13)$$

Third, we must know something about the collisional transition rates between the grains, and many models have been explored in the past, particularly those known as the step-ladder and the exponential models [72.R; 73.F; 77.T1]; I will confine myself almost exclusively in this discourse to the strong collision model, equation (2.26), and for the time being, at least, the apparent rate constant $\mu$ for the relaxation of the total internal energy can be regarded as an adjustable parameter.

It is important to understand two things about the graining procedure I have just described. One is the simple mechanical fact that equations (3.10)–(3.13) are only true if the grains are conceived to be very small, such that the Boltzmann factor $e^{-\varepsilon_{i,r}/kT}$ is constant for all $i$ within $r$; if not, the variation of the Boltzmann factor is taken into account by using equations (50)–(58) of [78.Y2]. The other is more fundamental: there is an intrinsic variability from one state to the next of the rate constants both for reactive decay and for collisional excitation or deexcitation; in choosing an average value for the whole grain, we are making an implicit assumption that molecules are in some way stirred among the states within the grain by some (as yet unknown) process whose rate is fast enough for it to pass unnoticed. I will return to a fuller discussion of this stirring process, i.e. randomisation, later on.

## 3.5 The stiffness problem

It should not be thought that the solution to the unimolecular rate problem is now at hand by numerical means, even though it might appear that all we require is a standard computer eigenvalue package and some imagination in guessing the elements of $Q$ and $D$. There remains a rather tricky little problem: most of the rate constants of the reactions shown in Figure 1.1 are of the order of $10^{-4} \, \text{s}^{-1}$ whereas typical predissociation rate constants can often be as large as $10^{13} \, \text{s}^{-1}$; perusal of Figure 3.1 shows that in a case like this, there will always be a factor of at least $10^{17}$ between the smallest and the largest eigenvalue. The set of differential equations (3.1) is said to be 'stiff', and this leads directly to a gross ill conditioning of the corresponding eigenvalue problem [52.C; 69.B; 71.M]: the net result of this ill conditioning is that the number of significant figures left in the computed value of $\gamma_0$ is roughly

$\log_{10}(\gamma_{max}/\gamma_0)$ – i.e. 17 in this example – fewer than the accuracy with which numbers are represented in the computer [75.P1]. A strategy for avoiding this problem has been worked out recently [82.P5].

# 4

## *The specific rate function $k(E)$ as an inverse Laplace transform*

At infinite pressure, as we argued in Chapter 1, the rate of replenishment of the reactive molecular states is so fast that even allowing for the decay processes, the populations of these states remain at their Boltzmann equilibrium values. Hence, we can write

$$k_\infty = \sum_i \tilde{n}_i d_i \tag{4.1a}$$

$$= \sum_r \beta_r d_r \tag{4.1b}$$

where the subscript $i$ refers to individual molecular states or $r$ refers to groups of states comprising the grains of equations (3.10)–(3.13); remember that the $d_i$, or $d_r$, are zero for the preponderance of the molecular states which are below the reaction threshold. If the grains span small adjacent segments of the energy spectrum, equation (4.1b) is equivalent to an integral

$$k_\infty = \{Q(T)\}^{-1} \int_0^\infty \rho(E) k(E) e^{-E/RT} dE \tag{4.2}$$

where $\rho(E)$ is the smoothed continuous approximation to the state degeneracies $P_r$, and $k(E)$ is the analogous approximation to the $d_r$; $\rho(E)$ is called the density of states at energy $E$, $k(E)$ is the specific rate function, and $R$ is the gas constant in appropriate energy units.[1] Notice that the division by the partition function does not appear formally in equations (4.1) since both the $\tilde{n}_i$ and the $\beta_r$ are normalised to unity for convenience in the eigenvalue methods described in the two preceding chapters. Also, as we noted above, $k(E)$ is zero below a certain threshold which we may call $E^*$, and thus equation (4.2) is, in practice, an integral only from $E^*$ to $\infty$. Moreover, since for even the simplest of molecules there are at least $10^4$ or $10^5$ states per wavenumber at[2] the energy $E^*$, we have no choice in

1 It is convenient at some point to slide over from the molecular units $\varepsilon$, $k$, as used in equations (3.10)–(3.13), to the practical units $E$, $R$, used by practising kineticists.
2 For a few molecules which span our general range of interest, e.g. $CO_2$, $CH_3NC$, cyclo-$C_4H_8$, $C_2F_6$, the densities of (rotation–vibration) states at the thermal thresholds are approximately $10^5$, $10^8$, $10^{15}$, $10^{21}$ states per wavenumber respectively.

any practical calculation but to evaluate $k_\infty$ over grains either as an integral, equation (4.2), or as a sum, equation (4.1b), which is computationally equivalent, rather than by equation (4.1a). Thus, our first requirement is to be able to calculate the density of states for any molecule of interest as a function of the energy $E$.

### 4.1 The partition function and the density of states

Looking back at equation (3.11), we see that it can be rewritten as

$$Q(T) = \int_0^\infty \rho(E) e^{-E/RT} dE \qquad (4.3)$$

In other words [38.B], $\rho(E)$ is the Laplace transform of $Q(T)$, whence if we know the behaviour of $Q(T)$ as a function of $T$ we can obtain $\rho(E)$ as an inverse Laplace transform [39.B]. The methods for performing this inversion are so well documented in the unimolecular reaction theory literature [39.B; 67.F; 69.F; 71.F; 72.R; 73.F] that there is little to be gained by repeating them here. However, the reader should note that the references cited do not constitute the last words on the subject: there have been significant improvements in computational methods in the last decade [77.Y2], and further improvements should still be anticipated. The density of states follows, then, once a set of molecular rotational constants and normal-mode vibration frequencies (together with their respective anharmonicities) has been assumed; these data are sufficient to define the partition function $Q(T)$, and the density of states $\rho(E)$ follows from the inverse Laplace transformation process. The numerical values, really, can only be considered as rather rough because of the imperfections of the molecular model, in particular the assumption of independent normal modes of vibration, a great paucity of vibrational anharmonicity data, the neglect of rotation–vibration interaction, and so on; moreover, their accuracy diminishes progressively as the energy increases.

In equation (4.2), we only need $\rho(E)$ at high energies where the density of states is high, the integral representation, equation (4.3), is adequate, and the Laplace inverse transform procedures are well defined. However, in the next section, the need will arise to know densities of states at rather low energies where the integral representation cannot be used: in these cases, the answer is obtained simply by enumerating the energies of all possible rotation–vibration states and then counting the number which fall into each energy grain. Such procedures go under the name of the direct count, and efficient algorithms were evolved by Beyer & Swinehart

[73.B1] and by Stein & Rabinovitch [73.S]; we have made some improvements [78.Y2] but, nevertheless, direct counts will remain costly in computing time and their use should be confined to as low an energy range as is practicable in the given problem.

The reader should also be aware of the effort expended in the past to try to obtain closed expressions for state densities: these are well summarised in [72.R; 73.F], though in all but the simplest cases, modern numerical methods will always be superior.

### 4.2 The rate law and the specific rate function

Examination of equation (4.2) shows that the way in which $k_\infty$ varies with temperature depends upon how both $k(E)$ and $\rho(E)$ vary with energy, although the dependence on $\rho(E)$ is the weaker because $Q(T)$ in the denominator is related to $\rho(E)$. For bimolecular reactions, Menzinger & Wolfgang [69.M3], and LeRoy [69.L2], have examined forms of rate law which arise when $k(E)$ is assumed to conform to certain simple functions of $E$, and a superficial conclusion would be that it is rather remarkable for chemical reactions in general to conform to the Arrhenius rate law. We saw in Chapter 1 that thermal unimolecular reactions do, in fact, conform very well to a strict Arrhenius rate law: the word *strict* is used here to distinguish the behaviour from a modified Arrhenius rate law, e.g. $k_\infty = B_\infty T e^{-\Theta_\infty/RT}$. This being so, then as was first pointed out by Slater [55.S], $k(E)$ must have a rather special form. In what follows, I will avoid the formal language of Laplace transforms as it tends to be somewhat confusing at first encounter; the reader is referred to a paper by Forst [72.F2] for a more elegant derivation. We rewrite equation (4.2) as

$$A_\infty e^{-E_\infty/RT} = \{Q(T)\}^{-1} \int_0^\infty \rho(E)k(E)e^{-E/RT}dE \qquad (4.4)$$

and multiply both sides by $Q(T)e^{E_\infty/RT}$, giving

$$A_\infty Q(T) = e^{E_\infty/RT} \int_0^\infty \rho(E)k(E)e^{-E/RT}dE \qquad (4.5)$$

Taking the constant term inside the integral on the right-hand side, equation (4.5) is the same as

$$A_\infty Q(T) = \int_{-E_\infty}^\infty \rho(E)k(E)e^{-(E-E_\infty)/RT}d(E-E_\infty) \qquad (4.6)$$

and if we now assume that the product $\rho(E)k(E)$ makes no contribution to the integral for $E < E_\infty$,

$$A_\infty Q(T) = \int_0^\infty \rho(E)k(E)e^{-(E-E_\infty)/RT}d(E-E_\infty) \qquad (4.7)$$

Thus, the product $A_\infty Q(T)$ is a Laplace transform of $\rho(E)k(E)$, but with a shift of $E_\infty$ in the variable of integration; extracting the inverse we have, by analogy with equation (4.3)

$$\rho(E)k(E) = A_\infty \rho(E - E_\infty) \qquad (4.8)$$

or, finally[3]

$$k(E) = A_\infty \rho(E - E_\infty)/\rho(E) \qquad (4.9)$$

Equation (4.9) is only defined for values of the energy $E > E_\infty$, and $k(E)$ is zero for all energies below $E_\infty$: thus, equation (4.2) becomes

$$k_\infty = \{Q(T)\}^{-1}A_\infty \int_{E_\infty}^\infty \rho(E - E_\infty)e^{-E/RT}dE \qquad (4.10)$$

and we have dispensed with much of the anticipated sensitivity of the calculated rate to inaccuracies in the state density since the latter is only needed at rather low energies where our approximate molecular model is fairly reasonable.

## 4.3 Criticisms of the inverse Laplace transform method

When Forst first suggested the use of equation (4.9) in unimolecular reaction calculations [72.F1; 72.F2], he did so (anticipating the likely criticisms) very cautiously. In retrospect, however, it is surprising that his proposal was not welcomed with more enthusiasm in view of the failing condition (to which I have already alluded in the Preface) of the transition state method. The advantage is that there is no longer any need to go through the elaborate procedures of RRKM theory in formulating an imaginary transition state – enough information pertaining to the nature of the reacting molecules is contained implicitly in the two observed quantities $A_\infty$ and $E_\infty$ [72.F3], and knowing them is sufficient to define an acceptable $k(E)$ function. In my view, Forst's suggestion has been met with undue suspicion (if not actual resistance) by the kinetics fraternity.

The major cause for concern has been that equation (4.9) is only true if the rate law is strict Arrhenius, i.e. $k_\infty = A_\infty e^{-E_\infty/RT}$ all the way from

---

3 Note that the derivation of equation (4.9) is equally valid if the molecular states are considered to be discrete [78.Y2], giving

$$k(E) = A_\infty P(E - E_\infty)/P(E) \qquad (4.9a)$$

absolute zero to infinite temperature; small, unmeasurable deviations from Arrhenius behaviour may give rise to unknown, and perhaps disastrous, errors in the derived $k(E)$ function [75.R; 77.R2; 78.B1]. A closely related concern is that specific rate functions derived in this manner will be inaccurate near threshold, and at very high energies, because of the limited range of temperatures over which the experimental rate law is known [76.R; 78.B1]. Yau and I have attempted to allay these suspicions in two quite different ways. I have collected together in Chapter 1 a series of observations which show that many thermal unimolecular reactions conform very closely indeed to a strict Arrhenius rate law at infinite pressure over exceedingly long temperature ranges; for such cases, equations (37)–(40) of [78.Y2] prove that equation (4.9) is a good approximation to the specific rate function. In another demonstration, we attempted to calculate realistic state-to-state rate constants for two simple unimolecular reactions, the thermal dissociations of nitrous oxide and of carbon dioxide [79.Y3; 79.Y4]; more details of one of these calculations are given later, in Chapter 6. From these state-to-state rate constants, it is possible to proceed in two ways. First, one can synthesise the $k(E)$ function directly by equation (3.12). Second, one can average these state-to-state rate constants over the Boltzmann distribution of reactant states for a series of temperatures to obtain the rate constants and thus the rate law; inversion of this rate law gives back a derived $k(E)$ function which agrees remarkably well with that calculated directly (see Figure 6.3 for this comparison). A similar comparison has been made recently by Forst [82.F1] with equally satisfactory results.

A second concern has been that energy of the first reactive state will not be exactly at an energy equal to the observed $E_\infty$ [74.T; 79.T3] and, especially if tunnelling phenomena are present, the molecular threshold may be considerably below the observed value of the activation energy; $k(E)$, however, according to equation (4.9) is not defined below $E_\infty$. Fortunately, reactive states near threshold make relatively little contribution to the total rate (see e.g. Figure 5.5) – unless, that is, we are dealing with the special case of a weak collision system near its low pressure limit; consequently, this imperfection will cause no significant error in the calculation of high pressure rate constants by equation (4.2) nor, especially if the system is a strong collision one, in the shape of the fall-off curves calculated according to the methods of Chapter 5. We should understand clearly the importance of the contribution of tunnelling to chemical reaction rates. In most practical cases, tunnelling makes only a very minor contribution to the reaction rate [73.A; 79.Y1; 82.C] and can be ignored. Where it can be important is when the reactant and product molecular states are connected by a tunnelling mechanism for a considerable range of energy below the classical energy threshold. If this is so, the rate law when measured at extremely low temperatures will be markedly non-Arrhenius: such a case is the thermal dissociation of nitrous oxide into a nitrogen molecule and an oxygen atom [82.F1]; it must also be pointed out that where the principal contribution to a reaction rate comes from tunnelling, accurate theoretical prediction of the reaction rate is a virtual impossibility because of the gross sensitivity of tunnelling rates to small imperfections in the assumed molecular potentials [79.P1; 80.E].

What if the rate law is non-Arrhenius? If the infinite pressure rate can be expressed with confidence in some other simple form, e.g. the modified Arrhenius equation

$$k_\infty = B_\infty T e^{-\Theta_\infty/RT} \tag{4.11}$$

the inverse Laplace transformation is equally valid, giving [78.Y2]

$$k(E) = R^{-1} B_\infty G(E - \Theta_\infty)/\rho(E) \tag{4.12}$$

which is remarkably like the usual specific rate function used in RRKM theory; $G(E)$ represents the sum of states up to energy $E$, i.e.

$$G(E) = \int_0^E \rho(E) \mathrm{d}E$$

instead of the density of states which appears in equation (4.9). Other simple rate laws can be handled similarly [79.F] and more complicated ones numerically [79.T1; 82.F1]; in fact, in the latter paper, Forst presents a tunnelling example which gives rise to a markedly non-Arrhenius rate law, and goes on to show that the inverse transform of that rate law corresponds satisfactorily with the known specific rate function.

I conclude that the objections to the idea of deriving the specific rate function $k(E)$, for use in describing thermal unimolecular reactions, as the inverse Laplace transform of the observed rate law are largely unfounded. Moreover, this procedure is superior to the RRKM procedure for calculating $k(E)$ on all counts: first, it is simpler; second, it gives a slightly better representation of the shape of the fall-off in strong collision systems (see Figures 2 and 9 of [78.Y2]); third, once the Arrhenius parameters are chosen, the RRKM $k(E)$ function is essentially invariant with respect to changes in the assumed transition state (see Figure 3 of [80.P1]).[4]

Finally, there have been some practical difficulties associated with the problem of calculating the densities of vibrational states at very low energies in equation (4.9) [72.F3], but these are readily overcome with a little computational effort [74.C]. In fact, the problem largely goes away when it is realised that it is the density of rotation–vibration states that is required in equation (4.9), not simply the density of vibrational states only. This conclusion follows inevitably from the fact that it is not even possible to find an adequate description of a simple diatomic dissociation reaction without taking full account of all the rotation–vibration states of the dissociating molecule [73.A; 75.P1; 79.Y1], and the same appears to be true for triatomics [76.P1]; moreover, experimental evidence of the crucial importance of rotation in the dissociation reactions of small polyatomic molecules is now becoming commonplace [82.P2].

4  Since the temperature variation of RRKM rate constants is modified Arrhenius, equation (4.11), to a very high degree of precision [59.S2], the appropriate $k(E)$ function is equation (4.12). However, instead of the experimental quantities $B_\infty$ and $\Theta_\infty$, 'theoretical' values according to a preconceived prescription are substituted and the $k(E)$ function is brought back to its proper mean numerical magnitude for the experimental temperature range by altering one or more of the molecular vibration frequencies; which ones are altered, it makes no essential difference [80.P1].

# 5

## Unimolecular fall-off in strong collision systems

The concept of a strong collision has been bound intimately with the development of the theory of unimolecular reactions ever since its inception in the 1920s. In 1927, Rice & Ramsperger introduced the approximation that an activated molecule would be deactivated whenever it suffered a collision, although they were well aware of the limitations in making such an assumption [27.R]; they supposed that it would be possible at a later stage to determine the extent to which this approximation was untrue, but I doubt if it was envisaged that it would take more than 50 years for us to achieve a realistic understanding of the problem. The principal reason for this delay has been the failure of kineticists to formulate a proper definition of a strong collision; the usual description has been a rather intuitive one [72.R; 73.F], generally adequate within the framework of a steady state treatment of the reaction process, but not always consistent with the principle of detailed balancing. Two useful (and equivalent) verbal definitions of a strong collision taken from the literature are as follows: a strong collision is one in which so much energy is transferred that the subsequent condition of the molecule may be chosen at random (with appropriate weighting factors for energy) from all its possible states [66.B2]; strong collisions are Markovian events whose outcome follows the probabilities given by the Boltzmann distribution, without reference to the initial conditions [77.T1]. These verbal definitions take on mathematical precision when we write [77.N]

$$q_{ij} = Z[M]\{Q(T)\}^{-1}e^{-\varepsilon_j/kT} \qquad (5.1)$$

where $Z[M]$ is the collision rate, $Q(T)$ is the partition function for the temperature $T$, and $\varepsilon_j$ is the energy of the final state $j$. Or, in the notation of Chapter 2 (and allowing for the fact that there is a degeneracy $g_j$ associated with states of energy $\varepsilon_j$), this is just

$$q_{ij} = Z[M]\tilde{n}_j \qquad (5.2)$$

where $\tilde{n}_j$ is the equilibrium population at energy $\varepsilon_j$ for the temperature in question. Substituting equation (5.2) into equation (2.4) gives exactly equation (2.26), i.e.

$$[Q]_{ij} = a_{ij} = \mu[(1 - \delta_{ij})\tilde{n}_i - \delta_{ij}(1 - \tilde{n}_i)] \tag{2.26}$$

except for the replacement of $Z[M]$ for $\mu$; thus, we can see immediately that the idea of pure exponential decay and the notion of strong collisions are, in fact, synonymous. At the same time, we can now give added mathematical precision to the ideas developed in Section 2.7, that a strong collision system is one in which there is no observable dispersion in the eigenvalues of the relaxation matrix and, conversely, a weak collision system will exhibit a significant dispersion of the eigenvalues.

### 5.1 Generalised strong collisions

We saw in Section 1.2(i) that the position of the fall-off on the pressure scale is determined by the competition between the rates of decay to products and of deexcitation of the reactive molecules. Also, we will see shortly that the calculation of the unimolecular rate constant through the use of equations (2.26), (3.4) and (4.9) gives a virtually exact representation of the shape of the fall-off, but that the pressure at which the fall-off occurs is only reproduced correctly if the relaxation rate $\mu$ is chosen to be about an order of magnitude smaller than the collision rate $Z[M]$. For this reason, Nordholm introduced the idea of an 'effective strong collision' whose rate constant is given by [78.N]

$$q_{ij} = \mu\tilde{n}_j \tag{5.3}$$

which leads precisely to equation (2.26); the term 'inefficient strong collision' has also been used [80.G2]. It is a little difficult to reconcile this concept with the verbal descriptions just quoted, especially that of Bunker [66.B2], but the interpretation is quite straightforward in the light of my new definition. We are facing here relaxation systems in which there are a great many eigenvalues, all very closely spaced. However, we have shown elsewhere in a model calculation of ultrasonic dispersion in hydrogen that it is barely possible to resolve experimentally a pair of modes of relaxation whose rate constants differ by a factor of ten [76.P2]; if there are more participating modes, resolution will become quite impossible. This lack of dispersion, then, between the eigenvalues can obviously give rise to the appearance of an unresolvable exponential decay in those systems which we regard as strong collision ones. The apparent rate constant for such a decay will be a suitable mean of the rates for all the modes contributing significantly to the relaxation, and if all the manifold of relaxation processes were to have rates ranging (say) from $0.01Z[M]$ to $Z[M]$, then we might then easily find $\mu = 0.1Z[M]$ as the weighted mean.

In the light of this discussion, I propose to generalise the concept of a strong collision to describe processes for which the relaxation eigenvalue dispersion is so small that the system appears to relax its *total* internal energy in a pure exponential manner.[1] Also, I will drop the cumbersome (and somewhat contradictory) adjectives 'effective' or 'inefficient', it being understood that the rate for such a process may well be less than the collision rate $Z[M]$ or even, as I will suggest in Chapter 7, very much greater than it; thus, from now on, when I will use the term strong collision, that will imply an unresolvable exponential relaxation with rate constant $\mu$, the fact that $\mu$ may never be equal to $Z[M]$ no longer being a cause for any discomfort.[2]

## 5.2 General properties of the eigenvalues

By the use of equation (2.10), we can rewrite equation (3.3) in the form

$$K = -E^{-\frac{1}{2}}[Q-D]E^{\frac{1}{2}} = [R+D] \tag{5.4}$$

whence the eigenvalue equation reads

$$[R+D-\gamma_i]\Psi_i = 0 \tag{5.5}$$

If we now choose the eigenvectors $\Psi_i$ such that $(\mathbf{S}_0, \Psi_i) = 1$, and remember that $\mathbf{S}_0 R = R\mathbf{S}_0 = 0$, then premultiplication of equation (5.5) by $\mathbf{S}_0$ yields very simply that

$$\gamma_i = (\mathbf{S}_0, D\Psi_i) \tag{5.6}$$

1 In practice, a polyatomic molecule exhibits only two internal relaxation rates, one associated with each of the rotational and vibrational relaxations. For a small molecule, the relaxation strength (i.e. the energy uptake [76.P2]) will be comparable for each type of process, and we may expect there to be an observable dispersion in the relaxation eigenvalues. On the other hand, for a large molecule, the vibrational contribution to the internal specific heat is dominant, with the rotational component almost completely submerged: under such conditions, we might not resolve the two sets of eigenvalues, giving what would appear to be strong collision behaviour, as I have defined it.

2 I had thought it advisable to invent another name for this property. Unfortunately, most synonyms for *strong* possess connotations of determinism, which is entirely inappropriate; adjectives like 'unbiased' could be used, but are cumbersome. 'Quasi-strong' is correct, but again cumbersome; also, I remember the time when we used the term 'quasi-unimolecular' for the reactions described in this book, so perhaps the qualification 'quasi-' would become redundant here too, before long.

or, more particularly, as is our main concern here,

$$\gamma_0 = (\mathbf{S}_0, D\mathbf{\Psi}_0) \tag{5.7}$$

This much is true, regardless of the nature of the collision matrix. Let us now substitute for $R$ the very simple form for the strong collision relaxation matrix given by

$$R = \mu(1 - p_0) \tag{2.29}$$

with $p_0$ designating the operator $\mathbf{S}_0(\mathbf{S}_0, \quad)$: the eigenvalue equation (5.5) then reads

$$[\mu - \gamma_i + D]\mathbf{\Psi}_i = \mu \mathbf{S}_0(\mathbf{S}_0, \mathbf{\Psi}_i) \tag{5.8}$$

after taking the $p_0$ term across to the right-hand side; notice that $[\mu - \gamma_i + D]$ is a diagonal matrix, as is also its inverse. Recalling the normalisation $(\mathbf{S}_0, \mathbf{\Psi}_i) = 1$, equation (5.8) can be rearranged to give the eigenvector

$$\mathbf{\Psi}_i = \mu[\mu - \gamma_i + D]^{-1}\mathbf{S}_0 \tag{5.9}$$

Finally, substitution of equation (5.9) into equation (5.6) gives us the eigenvalue

$$\gamma_i = \mu(\mathbf{S}_0, [\mu - \gamma_i + D]^{-1}D\mathbf{S}_0)$$
$$= \mu\sum_r [\mu - \gamma_i + d_r]^{-1}\beta_r d_r \tag{5.10}$$

The topology of these eigenvalues is consistent with the picture given in Figure 3.1. To a good degree of approximation, the $i$th eigenvalue is given by $(\mu + d_i)$: if there are $m$ identical entries of $d_i$ in $D$, then there are $(m-1)$ eigenvalues identically equal to $(\mu + d_i)$ and one which has the perturbed value

$$\gamma_i = \mu + d_i - \delta_i \tag{5.11}$$

Notice that there are very many zero entries in $D$ corresponding to all the unreactive grains below threshold, and so we have a very large number of eigenvalues of magnitude $\mu$. These perturbations $\delta_i$ are all small and in fact, because the trace of a matrix must equal the sum of its eigenvalues, we find that [79.P3; 81.V2]

$$\sum_i \delta_i = \gamma_0 \tag{5.12}$$

In principle, the $\gamma_i$ can be found by iterative solution of equation (5.10) but, except in the special case of $\gamma_0$, the process tends to be unstable [78.N]. There is a better procedure available, however, which allows one to iterate towards the $\delta_i$ of equation (5.11) without difficulty; it is described in the original paper [81.V2] and I will not reproduce it here.

## 5.3 The unimolecular rate constant

We are primarily interested in the smallest of the eigenvalues $\gamma_0$, for which we can write

$$\gamma_0 = \mu\sum_r[\mu - \gamma_0 + d_r]^{-1}\beta_r d_r \tag{5.13}$$

where $\beta_r$ is the equilibrium population of grain $r$ and $d_r$ is the specific decay rate constant for that grain. With the appropriate transcription of symbols, this is the same expression as that derived by Nordholm [78.N], equation 20. Experimentally, it is almost invariably the case that $\gamma_0 \ll \mu$: for example, in the methyl isocyanide or cyclopropane isomerisations, $\mu$ is typically $10^6$ s$^{-1}$ whereas $k_{uni}$ is of the order of $10^{-4}$ s$^{-1}$, and we can safely write for the rate constant an approximate eigenvalue

$$k_{uni} = \gamma_{0,ap} = \mu X = \mu\sum_r \beta_r d_r/(\mu + d_r) \tag{5.14}$$

which is the result of omitting $\gamma_0$ in the denominator of equation (5.13); if $\gamma_0$ is not small, then the rate constant is found exactly by iterative solution of equation (5.13). Conveniently, this expression possesses the same high pressure and low pressure limits as does the exact solution [81.V2], i.e.

$$k_{uni,\infty} = (S_0, DS_0) = \sum_r \beta_r d_r \tag{5.15}$$

and

$$k_{uni,0} = \mu\sum_r{}'\beta_r \tag{5.16}$$

where $\sum_r'$ represents a sum over only reactive grains. Also, in the intermediate pressure regime, equation (5.14) is a lower bound to the exact solution [81.V1]; notice that equation (5.15) implies $\Psi_0 \rightarrow S_0$ as $\mu \rightarrow \infty$. Notice also, again with the appropriate transcription of symbols, that equation (5.14) is the same expression as is used to calculate the rate constant in RRKM theory [78.N]. We have given elsewhere, and I will not rederive here, a rigorous upper bound expression [80.S; 81.V1]

$$\gamma_{ap} = \mu X/(1-X) \tag{5.17}$$

$X$ as defined in equation (5.14), which is identical with the very cumbersome combinatorial expression first discovered by Yau [78.Y1; 78.Y2].[3a,3b]

---

3 (a) Nordholm & Schranz [81.N] argue, correctly, that it is very little more effort to calculate the exact eigenvalue by equation (5.13) than it is to use the approximations, equations (5.14) or (5.17); however, the differences between the three results are minute except when $E_\infty/RT$ is small. They give five graphical examples which show these differences and at the same time illustrate the lower, upper bound properties of $\gamma_{0,ap}$ (equation (5.14)), $\gamma_{ap}$ (equation (5.17)) or, in their nomenclature, $\tilde{k}(\omega)$, $\hat{k}(\omega)$ respectively. Under these extreme conditions, it appears that equation (5.14) is a better approxi-

In equation (5.14), we have reached a very satisfying conclusion, that (by a rather remarkable chance)[3c] the usual RRKM formulation of the rate as a sum of independent Lindemann expressions is equivalent to the eigenvalue at large values of $E_\infty/RT$, and is therefore a reasonably correct expression for the rate constant; we will come back to a more detailed discussion of this equation later in this chapter.

### 5.4 The steady distribution

Once $\gamma_0$ is known, the eigenvector $\Psi_0$ can be generated exactly by using equation 15 of [81.V2]. Alternatively, an approximate eigenvector $\Psi_{ap}$ is given by

$$(\Psi_{ap})_r = \mu\beta_r^{\frac{1}{2}}(\mu+d_r)^{-1}/[1-\sum_r \beta_r d_r(\mu+d_r)^{-1}] \tag{5.18}$$

This can be verified easily, since $(S_0, D\Psi_{ap})$ returns the $\gamma_{ap}$ of equation (5.17), as it should. This result, then, gives us approximate values $\eta_r = (S_0)_r(\Psi_{ap})_r$ for the steady distribution, as described in Section 3.3: abbreviating the normalising constant in the denominator of equation (5.18) as $(1-\gamma_{ap}/\mu)$, approximately, we then find that

$$\eta_r = \beta_r/(1-\gamma_{ap}/\mu) \qquad \text{below threshold} \tag{5.19a}$$

$$\eta_r = \beta_r/[(\mu+d_r)(1-\gamma_{ap}/\mu)] \qquad \text{above threshold} \tag{5.19b}$$

Thus, unlike all model calculations cited in Section 3.3, the strong collision model gives the result that all states below threshold are in

mation than is equation (5.17): hence it would appear that equation (5.17), which we introduced with considerable enthusiasm, should follow our earlier combinatorial expression into the relative obscurity of the archives!

(b) We have alleged on several occasions that because equation (5.17) was an eigenvalue (or at least an excellent approximation to it), it therefore allowed properly for the mutual interference between the reactive channels, and would give superior results to the RRKM integration procedure [77.B1; 78.Y2; 79.P3; 80.S]; it is now clear that the procedures are equivalent, given the relative magnitudes of $\mu$ and $k_{uni}$ for which they are usually used, and that any difference between the reported calculations [78.Y2] arises solely from the use of different specific rate expressions.

(c) This absence of coupling between the reacting channels is a special consequence of the strong collision property: all depopulated grains are fed from the system at large at rates which will restore their equilibrium exactly simultaneously; no grain is favoured in this repopulation process, quite unlike the behaviour one would find in a step-ladder process. The reader might also find it curious that these expressions for $k_{uni}$ contain the quantities $\beta_r$, the *equilibrium* populations of the grains when, in fact, one is calculating a fallen-off rate constant $k_{uni}$ under what are clearly non-equilibrium conditions; this arises because the activation rate to each grain is proportional to $\beta_r$, as can be seen in prototype by examining equation (1.5). In some sense, the strong collision rate is a maximum: given a relaxation matrix $Q$ with eigenvalues $0, \lambda_i \leqslant \mu$, then $[\mu(1-p_0)+D] = S \geqslant [Q+D] = W$ and $\gamma_{0.S} \geqslant \gamma_{0.W}$.

equilibrium with each other;[3c] the slight apparent overpopulation of the unreactive states implicit in equation (5.19a) arises from the normalisation of the $\eta_r$ to unity, see [69.M2]. The reactive states, however, become more strongly depopulated with increasing $r$ because of the increase in $d_r$, according to equation (5.19b); also, all populations $\eta_r \to \beta_r$ as the relaxation rate $\mu \to \infty$ (i.e. as $p \to \infty$), as they should.

In fact, it is possible to do better than this, and to derive the whole evolution of the population distribution for all times, as given in equation 17 of [81.V2], but this result is probably of little more than academic interest in view of the unlikelihood of ever finding true strong collision behaviour.

## 5.5 Comparison with experiment

Two thermal unimolecular reactions may be regarded as the benchmarks against which to test the theory: these are both isomerisation reactions, of cyclopropane to give propylene and of methyl isocyanide to give methyl cyanide. It turns out, in fact, that methyl isocyanide is a borderline strong collision case, a point that we will discuss much more fully later, and so the comparisons made in this chapter will be limited to the cyclopropane reaction, wherever possible.

Figure 5.1 shows the observed rate constants for the isomerisation of cyclopropane to propylene [34.C; 45.C; 52.P; 53.P2; 54.S], measured at temperatures close to 490 °C, and all corrected to a reference temperature of 491.8 °C = 765 K through a knowledge of the activation energy; the data are plotted in the usual log–log fashion. The solid line in this figure is the theoretical strong collision prediction of the shape of the fall-off curve, obtained in the following manner.

The cyclopropane molecule is considered to be a collection of independent Morse oscillators whose observed fundamental vibration frequencies are given in Shimanouchi's tables [72.S1], and whose anharmonicity constants can be estimated as described either in [78.Y2] or in [80.P1]; this information, together with the external rotational constants of the molecule [78.Y2], is sufficient to yield the density of states $\rho(E)$ as a function of energy $E$. We then take the experimental values [61.F1; 78.Y2] for $E_\infty$ and $A_\infty$,[4] which enable us to calculate the inverse Laplace

---

4 It is clear from Figure 5.1 that there is excellent agreement between the temperature scales of [34.C], [45.C] and [53.P2]; these results do not, however, extrapolate well towards the infinite pressure value recommended in [61.F1], but to a figure about 4 or 5% higher, corresponding to a difference in temperature scales of about 0.8 K.

transform specific rate function

$$k(E) = A_\infty \rho(E - E_\infty)/\rho(E) \qquad (4.9)$$

Values of $k(E)$, or its grained equivalent $d_r$, are required up to sufficiently high an energy that the product $\beta_r d_r$ is small and can be neglected in the expression

$$k_{uni} = \mu \sum_r \beta_r d_r / (\mu + d_r) \qquad (5.14)$$

for the unimolecular rate constant; under conditions where equation (5.14) is a good approximation, this usually means considering values of $E$ up to a value $E_{max}$ about 30 kcal mol$^{-1}$ above $E_\infty$. The rate constant (5.14) is then evaluated by subdividing the range $E_\infty \leqslant E_r < E_{max}$, and for each value of the grain energy $E_r$ the grain population is calculated by equation (3.10) and the corresponding decay rate constant $d_r$ by equation (4.9). If equally spaced energy strips are used in this integration, then

Fig. 5.1. Comparison of the theoretical fall-off curve for the thermal isomerisation of cyclopropane at 765 K with the experimental results. The theoretical curve is calculated from the parameters listed in [78.Y2], except that the infinite pressure rate constant is taken to be $3.57 \times 10^{-4}$ s$^{-1}$ rather than $3.41 \times 10^{-4}$ s$^{-1}$ as recommended by Falconer, Hunter & Trotman-Dickenson [61.F1], see Footnote 4; the internal relaxation rate constant is $r_i = 5.50 \times 10^5$ Torr$^{-1}$ s$^{-1}$. The experimental data are those of Chambers & Kistiakowsky [34.C] (diamonds), Corner & Pease [45.C] (crosses), and of Pritchard, Sowden & Trotman-Dickenson [53.P2], also Appendix 2 (circles); the dotted line shows the high pressure limit of [61.F1].

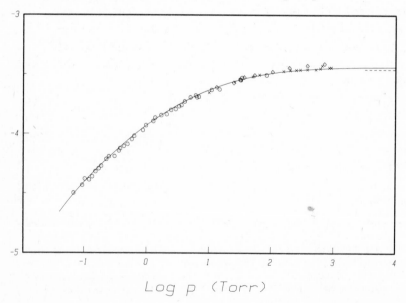

about 15–20 are sufficient to give an acceptably accurate rate constant; fewer strips would be required, however, if a Laguerre quadrature [64.H; 67.H] were to be used. Finally, it is necessary to choose appropriate values for the (first order) strong collision relaxation rate constant

$$\mu = pr_i \qquad (5.20)$$

where $r_i$ is the (second order) rate constant for the internal relaxation and $p$ is the pressure. Varying $r_i$ causes the calculated fall-off curve to be moved horizontally without altering its shape: increasing $r_i$ moves the curve to the left; conversely, decreasing $r_i$ moves the curve to the right; the virtually perfect agreement between strong collision theory and experiment is obtained by choosing $r_i = 5.5 \times 10^5$ Torr$^{-1}$ s$^{-1}$, which corresponds to a value of $\mu$ of about one-tenth of the hard sphere collision rate $Z[M]$. Similar, almost equally satisfying, agreements between experiment and strong collision theory can be seen elsewhere in the literature, particularly [78.Y2]; the apparent internal relaxation constants required to position the fall-off correctly are shown in Table 5.1, and, in all cases, the apparent relaxation rate is about an order of magnitude below the hard sphere collision rate.

We can carry this comparison with experiment a little further by measuring the rate of the cyclopropane isomerisation at sufficiently low pressures that the mean free path exceeds the diameter of the vessel [63.K2].[5] A 1 l spherical flask was used, and the results are shown in Figure 5.2: the solid theoretical line is calculated on the assumption that the internal energy relaxation rate is given by

$$\mu = pr_i + (\hat{c}/4) \times (S/V) \qquad (5.21)$$

where $S/V$ is the surface to volume ratio of the vessel, $\hat{c}$ is the mean molecular speed at the temperature of the gas, and $r_i$ has the value listed in Table 5.1. Notice that the rate of collision with the wall is unambiguous, depending only upon the mass of the molecule, the temperature, and the pressure [64.E]; thus, the only new assumption made in writing equation (5.21) is that the wall has unit collision efficiency for the relaxation of the

5 There has been some sporadic warfare in the literature over the years concerning the integrity of these experiments [69.T; 78.Y2]: the question at issue is whether a 4 cm diameter trap, permanently connected to the reaction vessel by about 15 cm of 2 cm diameter tubing, and permanently maintained at $-80°$C [63.K1] would keep the desorbed water in the reaction mixture at its equilibrium vapour pressure of $4 \times 10^{-4}$ Torr, or not. Since the efficiency of water in maintaining the fallen rate is about 0.8 that of cyclopropane [53.P2], the expected increase in rate is too small to account for the magnitude of the observed rate under the surface activated conditions, as is shown in Figure 5.2.

Table 5.1. *Apparent internal relaxation rate constants for strong collision reactions*

| Molecule | $T$ (K) | $m$ (g mol$^{-1}$) | $\sigma$ Å | $Z$ Torr$^{-1}$ s$^{-1}$ | $r_i$ | $r_i/Z$ |
|---|---|---|---|---|---|---|
| $CH_3NC$ | 503.6 | 41.0 | 4.5 | $8.80 \times 10^6$ | $1.20 \times 10^6$ | 0.14 |
| $CD_3NC$ | 503.6 | 44.0 | 4.5 | $8.49 \times 10^6$ | $1.30 \times 10^6$ | 0.15 |
| $C_2H_5NC$ | 504.1 | 55.1 | 5.5 | $1.13 \times 10^7$ | $1.32 \times 10^6$ | 0.12 |
| $C_3H_6$ | 765.0 | 42.1 | 4.0 | $5.55 \times 10^6$ | $5.50 \times 10^5$ | 0.10 |
| $C_4H_8$ | 721.5 | 56.1 | 4.5 | $6.29 \times 10^6$ | $5.65 \times 10^5$ | 0.09 |
| $C_2H_5Cl$ | 794.2 | 64.5 | 4.5 | $5.59 \times 10^6$ | $5.80 \times 10^5$ | 0.10 |

To convert from units of cm$^3$ molecule$^{-1}$ s$^{-1}$ to Torr$^{-1}$ s$^{-1}$, multiply by $9.658 \times 10^{18}/T$.

internal energy, compared with an apparent efficiency of only about one-tenth for the gas–gas collisions. Another set of points in Figure 5.2 shows the effect of packing the reaction vessel with short lengths of glass tubing: the limiting low pressure rate increased, as expected, but because the packing was of an irregular geometry, it is not easy to predict what this new rate should have been.[6]

## 5.6 The shape and position of the fall-off

We now have an extremely simple procedure by which to calculate the shape of the fall-off of the unimolecular rate constant in the strong collision approximation; the position of the fall-off can be varied at will simply by changing the (unknown) rate constant $r_i$ for the relaxation of the internal energy of the reacting molecule. It should therefore be possible to provide a reasonable explanation of the variation in both the position and the shape of the fall-off (especially the latter), such as is shown in Figure 1.1, with varying complexity of the reacting molecule. There are only two basic ingredients in the calculation, equations (4.9) and (5.14).

Let us assume for a moment that all the $d_r$ in equation (5.14) can be taken as a constant $d$: we then have

$$k_{uni} = n'd\mu/(\mu+d) = n'd[pr_i/(pr_i+d)] \tag{5.22}$$

---

6 Comparison of the model calculation shown in Figure 1.3 for a 16 ml sphere suggests that a somewhat larger increase in rate than this might have been expected upon packing the vessel completely with 4 cm lengths of 1 cm diameter tubing, despite the randomness of their disposition within the flask. This would be a relatively easy experiment to repeat with modern equipment.

where $n' = \Sigma_r' \beta_r$. This equation is of the strict Lindemann form:[7] when plotted in the usual way (log $k_{uni}$ v. log $p$) it is a horizontal straight line of value $n'd$ at high values of $\mu$, and at very low values of $\mu$ it is again a straight line of slope one with magnitude $\mu n'$. Also, the point of maximum curvature occurs when $\mu = d$, i.e. $p = d/r_i$; this, as we saw earlier, is $p_{\frac{1}{2}}$, the so-called half-pressure. Thus, the position of this fall-off curve is defined unambiguously by the ratio between the mean decay rate constant $d$ and the mean relaxation rate constant $r_i$. But the $d_r$ of equation (4.9) are not constant: they increase strongly with increasing energy above reaction threshold. To understand what happens, rewrite equation (5.14) as

Fig. 5.2. Comparison of the theoretical fall-off curve for the thermal isomerisation of cyclopropane in a 1 l sphere at 765 K with the experimental results from [63.K2], also Appendix 2 (circles). The dotted curve is that corresponding to an infinitely large vessel, as shown in Fig. 5.1, and the much fainter dotted curve shows by how much the rate would be increased by the water vapour which is known to be present. The crosses are experimental results for a packed reaction vessel, and the dashed line is a fit to these data, which corresponds to a sphere of about 16 ml.

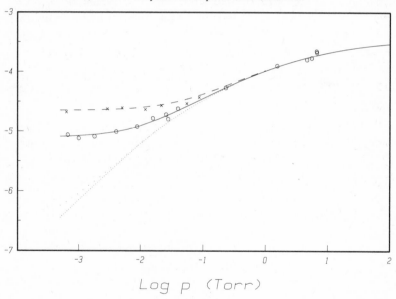

Log p (Torr)

7 Notice that the sufficient condition for the occurrence of strict Lindemann behaviour, given a strong collision relaxation matrix, is that the $d_r$ are constant; alternatively, if there is only one grain having a non-zero value of $d$, the rate is also strict Lindemann, *regardless* of the form of the assumed relaxation matrix [81.V1]. In the past, it had often been assumed that strict Lindemann behaviour was a strong collision property only, but we now know that near-Lindemann behaviour can often occur in weak collision systems at high temperature, see Chapter 8.

$$k_{uni} = \sum_r \beta_r d_r [\mu/(\mu + d_r)] = \sum_r \beta_r d_r [pr_i/(pr_i + d_r)] \qquad (5.23)$$

and we see that $k_{uni}$ is a sum of Lindemann curves, but each curve is centred about a different pressure $p_{\frac{1}{2},r} = d_r/r_i$. To proceed further, we must now consider the functional forms of $\beta_r$ and $d_r$ in more detail. The quantities $\beta_r$ are simply the Boltzmann equilibrium populations for the grains $r$; these decrease very strongly with increasing energy so that the values shown for cyclopropane in Figure 5.3 are plotted in logarithmic form. Likewise, Figure 5.4 shows, again in logarithmic form, the manner in which the $d_r$ increase with energy above threshold; we will return in the next chapter to a more detailed examination of the form of the specific rate function. From these two functions, we can then construct the products $\beta_r d_r$ as shown in Figure 5.5: these products increase to a broad maximum with increasing energy and then fall away quite strongly at high energies. Let us now dissect equation (5.23) into the $\beta_r d_r$ terms and the terms in square brackets: the latter we may regard as primitive Lindemann shapes, each centred about its own value of $p_{\frac{1}{2},r}$. Consequently, whenever the magnitude of the product $\beta_r d_r$ is significant, there is a primitive Lindemann contribution to $k_{uni}$ centred about the pressure $d_r/r_i$; thus, the broader the peak in the $\beta_r d_r$ curve, the broader is the dispersion of the centres $p_{\frac{1}{2},r}$ of these contributing Lindemann forms and so the longer the range of pressures it takes for the sum, $k_{uni}$, to change

Fig. 5.3. Equilibrium populations, $\beta_r \equiv \tilde{n}(E)$, for cyclopropane (solid line), and (dashed line) for a hypothetical molecule of similar structure but with fewer internal degrees of freedom (see text), as a function of energy above threshold.

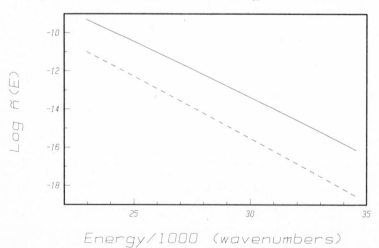

Log $\tilde{n}(E)$

Energy/1000 (wavenumbers)

Fig. 5.4. Specific rate functions, $d_r \equiv k(E)$, for the two molecules of Fig. 5.3, as a function of energy above threshold. Note the minor irregularities at low energies, due to the changeover from direct count to approximate methods for calculating the state densities.

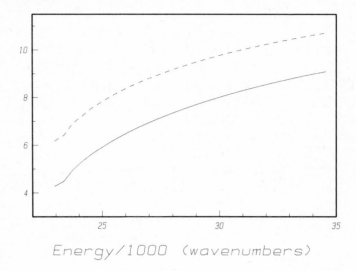

Energy/1000 (wavenumbers)

Fig. 5.5. Variation of the product $\beta_r d_r$ with energy above threshold for the two molecules of Fig. 5.3; the dotted line shows the hypothetical variation of $\beta_r d_r$ for cyclopropane if all $d_r = d = \text{constant} = 2.72 \times 10^5 \text{ s}^{-1}$, i.e. the Lindemann approximation.

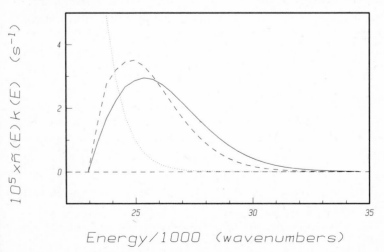

Energy/1000 (wavenumbers)

over from a slope of one to a slope of zero.[8] And of course, just as the position of the fall-off of the strict Lindemann form was defined by the ratio of $d/r_i$, so in this case the fall-off is centred roughly about the pressure $d_{r,av}/r_i$, where $d_{r,av}$ is the mean value of $d_r$ derivable, in principle, from Figure 5.5. The net results of these effects are to be seen in Figure 5.6, where the leftmost solid line is the true fall-off curve for cyclopropane, and the dotted line is the strict Lindemann form having the same limiting values of $k_{uni,0}$ and $k_{uni,\infty}$; the dispersion of the contributing $d_r$, shown in Figure 5.5, causes a marked moderation of the sharpness of the fall-off curve compared with the case where $d_r$ is constant.

So what happens if we change our consideration to a molecule of different complexity? In practice, there are many variables which complicate the analysis, for not only will the $\beta_r$ and $d_r$ change, but $\mu$, $E_\infty$ and $A_\infty$ will also be different. Let us imagine a hypothetical molecule $C_3D_3$ which possesses the same internal relaxation rate constant as does cyclopropane, and which reacts to form some product with the same values of $E_\infty$ and of $A_\infty$. We will also assume that it has the same two moments of inertia as does cyclopropane, so that the only thing different about it is its vibrational frequencies: it has 12 normal modes of vibration instead of 21, and for the purposes of this illustration, I have simply made an arbitrary deletion of nine of the original modes of the cyclopropane molecule.

The Boltzmann populations $\beta_r$ of the grains are shown as the dashed line in Figure 5.3; these populations are much lower than for the cyclopropane case because the simpler molecule has (relatively) a much lower density of states at the energies of interest. At the same time, the corresponding values of $d_r$, shown in Figure 5.4, are markedly increased, and we need to understand why this is so. Equation (4.9) admits the following interpretation: the ratio $\rho(E-E_\infty)/\rho(E)$ is the fraction of the total states at energy $E$ which have an amount of energy $E_\infty$, exactly, 'locked up' in some unspecified fashion, and all such states decompose to products with an effective rate constant $A_\infty$. At low energies, there are relatively few vibrational states in either molecule, and the density of states is essentially that of the rotational states, which is the same for both molecules: thus, since our hypothetical molecule has far fewer states at high energies $E$ but roughly the same number of states at low energies $(E-E_\infty)$, the ratio $\rho(E-E_\infty)/\rho(E)$ is much larger and so, therefore, is

8 Elsewhere, we have given an alternative, but equivalent, explanation in terms of the interaction of two bundles of eigenvalues, one bundle of magnitude $\mu$, and the other with magnitudes $d_r$ and a dispersion defined by the distribution $\beta_r d_r$ [81.V1].

its specific decomposition rate $k(E)$, or $d_r$. With the same $r_i$, as we have assumed, this immediately places the position of the fall-off at much higher pressures (since $d_{r,av}/r_i$ has to be much larger, regardless of the distribution effect from the product $\beta_r d_r$), and this is seen to be so in Figure 5.6, the rightmost solid curve.

We now have to examine the distribution function $\beta_r d_r$, which is shown as the dashed curve in Figure 5.5; it lies somewhat to the left of the cyclopropane distribution curve, and exhibits a lesser dispersion. This happens because in the simpler molecule, the Boltzmann populations $\beta_r$ fall off about five times faster than in cyclopropane over the energy range from $23\,000\ \text{cm}^{-1}$ to $34\,000\ \text{cm}^{-1}$ (although this is compensated to some extent by a slightly faster increase in the specific rate function for the smaller molecule). Thus, the narrowing of the dispersion of the function $\beta_r d_r$ is *primarily* a consequence of the fact that the Boltzmann distribution function falls more rapidly for simpler molecules, and this is therefore the direct cause of the sharpening up of the fall-off curve as the reacting molecule becomes simpler; the extent of this sharpening becomes apparent when the fall-off curve for our imaginary molecule is shifted

Fig. 5.6. Comparison of the fall-off curves for cyclopropane and for a hypothetical molecule with fewer internal degrees of freedom (see text), shown as two solid lines, left and right respectively. The dashed curve is that for the hypothetical molecule, but shifted to the left so as to show the difference in shape between the two curves; the dotted line is the strict Lindemann shape.

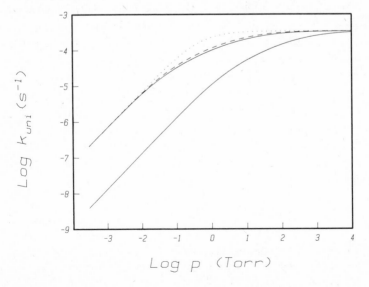

leftwards so that its low pressure limiting rate coincides with that of cyclopropane.

This explanation for the positions and the shapes of unimolecular fall-off curves is very straightforward, and considerably less elegant than that which is usually offered. For the best part of 50 years now, kineticists have used a shape parameter $s$, generally known as the Kassel $s$, to describe in a semiquantitative manner the degree of curvature of the fall-off. Throughout all these years, the appropriate numerical value of $s$ has been thought to be about half of the number of vibrational degrees of freedom in the reactant molecule, but, as we have shown recently [81.V1], this correlation is an accidental one, and has no firm physical basis. We are left in the following position: inspection of equation (5.22) shows that as $\mu$ changes in magnitude from $0.01d$ to $100d$, the strict Lindemann curve changes (for all practical purposes) from second order to first order behaviour; in other words, the range of the curvature in the strict Lindemann case is four decades in the pressure, whereas for cyclopropane the curved portion of the fall-off extends over more like six decades in the pressure. The Kassel $s$ has been an extremely useful descriptor for these subtle changes in shape, but the connection between $s$ and the physical makeup of the molecule is, at best tenuous, at worst without foundation; it should, however, be possible to generate empirical broadening factors for strong collision fall-off curves, rather along the lines given by Troe for weak collision cases [79.T2], but it does not seem at the moment that they will possess the same intuitive quality nor be as easy to use as the Kassel $s$. An alternative approach might be to construct a generalised Lindemann expression containing an additional parameter which governs the range of pressure over which the transition from second order to first order behaviour occurs.

Small changes in the shape and position of the fall-off occur if the temperature is altered, and we will deal with these in the next section. We will not scrutinise the other possible permutations:[9] for example, with a given molecular complexity and fixed $E_\infty$, the position of the fall-off is fixed by the ratio of $A_\infty/r_i$, and so on; on the other hand, if $E_\infty$ and $A_\infty$ are changed simultaneously so as to keep $k_\infty$ unchanged, more-complex shifts occur, see e.g. [62.W].

---

9 The student who wishes to study these in more detail may do so quite easily with the help of the computer program given at the end of this book.

## 5.7 The effect of temperature

Inspection of equation (5.14) reveals that a change in temperature will affect the unimolecular rate for a strong collision system in two ways: the dominant effect will be through the change in equilibrium populations $\beta_r$, with a much smaller contribution arising from the temperature dependence, so far unknown, of the internal relaxation rate constant $\mu$. Although we do not know the exact nature of this relaxation process, it is safe to assume that, like most other internal relaxation processes, it has a rather small temperature coefficient which we can ignore for the time being. As the temperature is raised, the Boltzmann populations $\beta_r$ of the excited states increase strongly at the expense of the low-lying ones, and the rate increases strongly too (at the low pressure limit, we have $k_{uni,0} = \mu \Sigma_r' \beta_r$, and at the high pressure limit, $k_{uni,\infty} = \Sigma_r \beta_r d_r$). As a result, the dispersion of the function $\beta_r d_r$, as illustrated in Figure 5.5, becomes broader, and therefore the shape of the fall-off curve itself becomes broader; at the same time, the peak in the $\beta_r d_r$ function moves to higher energies, which means that the mean value of $d_r$, i.e. $k(E)$, for the reaction increases, and so the centre of the fall-off curve moves to higher pressures, determined by the condition of the equality between the mean specific reaction rate and the internal relaxation rate.

Significant segments of the fall-off curves for the thermal isomerisation of methyl isocyanide have been measured at four temperatures [62.S; 66.F], and the expected behaviour is clearly demonstrated: Figure 4 of [66.F] shows a plot, in the low pressure region, of log $k/k_\infty$ v. log $p$ for all four temperatures and, despite the experimental scatter, the progression to higher pressures of the fall-off as the temperature rises from 473 to 553 K is quite unambiguous; a similar diagram is given in [60.C] for the isomerisation of methylcyclopropane at 720 and 763 K. For the methyl isocyanide experiments reported in [62.S], (interpolated) values of $p_{\frac{1}{2}}$ are 31, 40, and 47 Torr at 473, 504, and 533 K respectively;[10] likewise, for the thermal isomerisation of cyclopropane, the values of $p_{\frac{1}{2}}$ are 1.5 [60.S], 4.5 [53.P2], and 17 Torr [82.F2] at 718, 765, and 897 K respectively.

Changes in shape of the fall-off curves over the observed temperature ranges are much less dramatic, and the effects are shown in Figure 5.7 for both the methyl isocyanide and cyclopropane reactions, in the strong

---

10 The half-pressures quoted here for both reactions are for the strong collision calculation: as explained in Chapter 7, the methyl isocyanide reaction does not conform exactly to the strong collision properties and the observed values of the half-pressure are actually some 20 Torr higher, see Figure 7.3.

collision approximation; such variations have been explored numerically in considerable detail elsewhere [65.P; 72.S2], and over much wider temperature ranges. The experimental data conform to the theoretical shapes shown in Figure 5.7 remarkably well, as can be seen for cyclopropane in Figure 5.8, where the range of temperature is almost 200 K; there is, however, a slight suspicion that, at the highest of these temperatures, the fall-off may be a trifle sharper than that predicted by the strong collision model, a point which is worthy of further experiment.

An alternative way of treating this problem, one which does not separate the variation in shape of the fall-off from its shift with temperature, is to examine the variation of the activation energy for the reaction with changes in pressure. Here, the use of the term 'activation energy' implies nothing more than a convenient way of labelling the magnitude of the variation in rate constant with temperature, viz.

$$E_p = -R\mathrm{d}\ln k_{\mathrm{uni},p}/\mathrm{d}(1/T) \qquad (5.24)$$

For any unimolecular reaction at its high pressure limit, we can, following Tolman [20.T], give a physical interpretation of this experimentally determined quantity. We rewrite equation (4.1b) in the form

Fig. 5.7. Variation of the shape of the strong collision fall-off curve with temperature: rightmost pair, methyl isocyanide at 473 and 533 K; leftmost pair, cyclopropane at 718 and 897 K. In each case, the low temperature curve is shifted so as to make each member of the pair coincide at both limits.

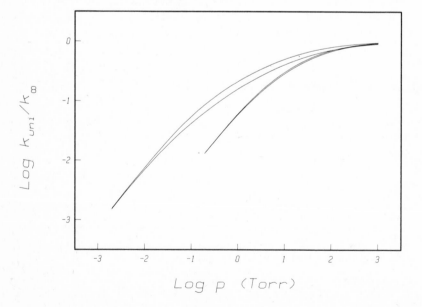

$$k_{\text{uni},\infty} = \frac{\Sigma_r g_r d_r e^{-E_r/RT}}{\Sigma_r g_r e^{-E_r/RT}} \qquad (5.25)$$

then, differentiating equation (5.25) according to equation (5.24), we get

$$E_\infty = \frac{\Sigma_r g_r d_r e^{-E_r/RT} E_r}{\Sigma_r g_r d_r e^{-E_r/RT}} - \frac{\Sigma_r g_r e^{-E_r/RT} E_r}{\Sigma_r g_r e^{-E_r/RT}} \qquad (5.26)$$

$$= \bar{\bar{\varepsilon}}_r - \bar{\varepsilon}_r$$

where $\bar{\varepsilon}_r$ is the mean energy of all molecules in the system, and $\bar{\bar{\varepsilon}}_r$ is the mean energy carried away from the assembly by those molecules which react. For the strong collision case, we can make a similar derivation for the low pressure rate, $k_{\text{uni},0}$: we pick up equation (5.16) and differentiate it in like manner, yielding that

$$E_0 = E_\mu + \varepsilon'_r - \bar{\varepsilon}_r \qquad (5.27)$$

where $E_\mu$ is the Arrhenius temperature coefficient of the internal relaxation rate, $\varepsilon'_r$ is the mean energy (at thermal equilibrium) of all states above threshold, and $\bar{\varepsilon}_r$ is, as before, the mean energy (again at thermal equilibrium) of all molecules in the system. Since $\bar{\bar{\varepsilon}}_r > \varepsilon'_r$, $E_\infty > E_0$ if $E_\mu$ can be assumed to be negligible; moreover, the more complicated the

Fig. 5.8. Comparison of calculated strong collision fall-off curves with experimental data for cyclopropane at three temperatures, 718 K [58.S; 60.S] (crosses), 765 K [53.P2] (squares), and 897 K [82.F2] (triangles).

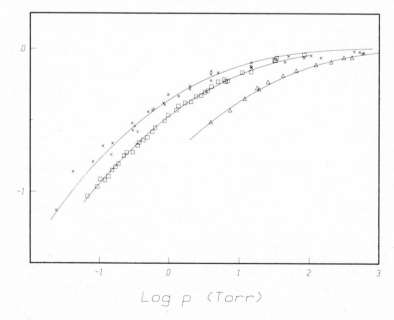

molecule, the greater is the difference $(E_\infty - E_0)$, as can readily be seen from a study of Figures 16–18 of [78.Y2] and the discussion thereof.[11,12]

At intermediate pressures, no such simple decompositions as equations (5.26) and (5.27) are possible [32.K], but it requires only a trivial extension of our computational procedures to obtain $E_p$ numerically, i.e.

$$E_p(\beta) = \{\Delta\beta\}^{-1}\ln[k_{\mathrm{uni},p}(\beta)/k_{\mathrm{uni},p}(\beta + \Delta\beta)] \qquad (5.28)$$

where $\beta = 1/RT$. The results of using equation (5.28) together with equation (5.14) are compared with two sets of experimental values [62.S; 76.C] for the methyl isocyanide reaction in Figure 5.9, the solid line corresponding to the assumption that the internal relaxation rate is

Fig. 5.9. Variation of the activation energy for the thermal isomerisation of methyl isocyanide as a function of pressure. Experimental data: squares from Schneider & Rabinovitch [62.S], triangles from Collister & Pritchard [76.C]. The solid line assumes that $\mu$ is independent of temperature; the dotted line assumes that $\mu$ increases slightly with temperature, with an Arrhenius temperature coefficient of 0.25 kcal mol$^{-1}$.

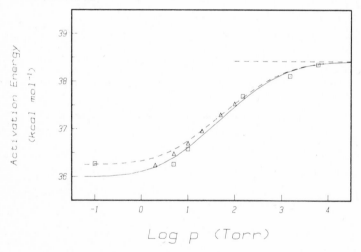

11  In the case of cyclopropane, $E_0$ was assumed [78.Y2] to be the same as the activation energy measured in the wall activation regime, see Figure 5.2; this presupposes that the temperature coefficient for the wall activation processes is also negligible, which undoubtedly is true. Notice that the activation energies for the wall sustained reaction drop markedly with increasing temperature, from about 58 kcal mol$^{-1}$ near 700 K [63.K2] to around 45 kcal mol$^{-1}$ near 1000 K [82.Y1], as required by equation (5.27).

12  It should be understood that there is not necessarily any direct connection between the threshold energy, $E^*$, for the reaction and any measured Arrhenius temperature coefficient; *however*, as shown here and in the preceding chapter, the *assumption* of strict Arrhenius behaviour with the threshold taken as $E_\infty$ gives an excellent representation of the fall-off shape *for strong collision reactions*.

independent of temperature; the agreement between theory and experiment is excellent. The dotted line shows the effect of allowing a small temperature dependence ($0.25 \, \text{kcal mol}^{-1}$) for the internal relaxation rate $\mu$: clearly, the temperature dependence of $\mu$ must be very small, but these experimental results do not contain enough information for us to be able to assess the precise magnitude of the variation. Correspondingly, neither set of data [62.S; 76.C] has sufficient precision for an unambiguous determination of the temperature coefficient $E_\mu$ to be made from a comparison of the fall-off curves themselves; the same is true of the temperature coefficient of $\mu$ for the cyclopropane isomerisation reaction, despite the much wider temperature range of the three independent determinations [53.P2; 60.S; 82.F2].[13]

We have already summarised, in Chapter 1, the remainder of the experimental measurements of the variation of activation energy with pressure, including some weak collision examples; a proper treatment of the latter must await a better understanding of the nature of the bottleneck phenomena (as described in Chapter 8) which underlie the weak collision behaviour.

## 5.8 Simultaneous thermal reactions

Except for a diatomic molecule, there is no unique way in which a molecule may dissociate, or otherwise rearrange itself, when heated. For very simple molecules, the thresholds in energy for the different possible processes tend to be widely spaced and so, in general, only the one reaction requiring the least energy of activation is observed. As the unimolecular reactant molecule increases in complexity, the number of possible pathways multiplies enormously, and it becomes quite common for there to be two, or three (or even more) distinct sets of products formed from one reactant species, all having rather similar threshold energies, see Table 7.1 of [72.R], for example; under these conditions, identification and quantification of the various products are more readily achieved.

Suppose we have a reactant R which decomposes to products P above one threshold; then, above a somewhat higher threshold, R may decompose to give a new set of products Q in addition to the products P. Remembering that we are concerned with strong collision systems (because of the molecular complexity), then it is clear that the products P and Q are formed in competition with each other, grain-by-grain. As we

13 We will discuss this problem in a little more detail in Chapter 9.

follow the reaction down from high pressures into the fall-off region, the populations of the states above the lower threshold begin to fall, according to equation (5.19b), as the result of the reaction to form P; when the second threshold is reached, this depletion of population continues, and is superimposed on the depletion that would have been caused by the reaction to form Q. Thus, it is clear that the two reactions interfere with each other at any pressure in the fall-off region. Most of the reactive flux arises from states of the reactant which are not very far above threshold (see Figure 5.5 again), and this is especially so as the pressure is reduced. Hence, if there is a sufficient difference in the threshold energies, the formation of Q will have very little effect on the rate of formation of P, but the rate of formation of Q will be subject to strong interference by the reaction to form P. Consequently, the fall-off behaviour of the reaction to form P will be only a little different from that which would occur if there were no other reactive channels available; on the other hand, the reaction to form Q will exhibit a much sharper fall-off, characteristic of a much smaller reactant molecule, and its half-pressure is moved correspondingly to a higher value [77.B1]. This observation that the channel with the larger energy requirement falls off faster was first made by Chesick in a study of the thermal decomposition of methyl-cyclopropane [60.C].

The strong collision master equation is readily solvable for the case of competing reactions with separate thresholds [81.V2]: the result takes the form of generalisation of equation (5.17) and, neglecting the denominator as before, we have

$$\gamma_{\mathrm{ap},i} = \mu\sum_r \beta_r d_{r,i}/(\mu + d_r) \tag{5.29}$$

where $d_{r,i}$ is the specific decomposition rate for the $i$th channel, and $d_r = \Sigma_i d_{r,i}$. With the appropriate transcription of symbols, this formula is the same as that given by King, Golden, Spokes & Benson [71.K2], as a result of intuitive reasoning; equation (5.29), however, is one of a series of bounds to the exact eigenvalue, with known properties [81.V2], (see Footnote 3a earlier in this chapter).

There are not many sets of experimental data against which to test equation (5.29). An interesting example is that of the thermal decomposition of monofluorocyclopropane, which gives rise to four distinct products, *cis-* and *trans-*1-fluoropropene, 2-fluoropropene, and 3-fluoropropene [64.C]. The two curves drawn in Figure 5.10 show the result of applying equation (5.29) to this set of reactions; only the experimental rate constants for the production of *trans-*1-fluoropropene and of

2-fluoropropene are shown for simplicity.[14] I want to call attention to two points: one, that the second reaction, which has an activation energy of about 65 kcal mol$^{-1}$, falls off rather more sharply than does the other whose activation energy is nearer to 61 kcal mol$^{-1}$; and two, that a single value of the internal relaxation rate constant brings about a reasonable coincidence of both sets of data with the theoretical fall-off curves. I should emphasise that this was a very difficult experiment, as reference to the original paper will show immediately, and the correctness of the

Fig. 5.10. Comparison of fall-off curves for two of the four reaction products in the thermal isomerisation of monofluorocyclopropane, in the strong collision approximation. The upper theoretical curve corresponds to the rate of formation of *trans*-1-fluoropropene, and the lower one to that of 2-fluoropropene. The points are the experimental results of Casas, Kerr & Trotman-Dickenson [64.C]; see Footnote 14 also; the position of these curves is determined by an assumed internal relaxation rate constant $r_i = 3 \times 10^5$ Torr$^{-1}$ s$^{-1}$.

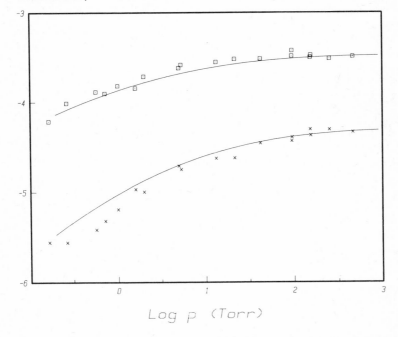

14 I do not show the remaining experimental points (nor the corresponding theoretical lines, which lie in between the pair shown) mainly for the sake of clarity. The points for 3-fluoropropene scatter about the theoretical line, but so widely that they intermingle on occasion with the data above and below. Those for *cis*-1-fluoropropene also fall near the theoretical curve, but somewhat to the left of it; there are several reasons why this might happen, one of which is suggested in [78.F1], but the experiment should be repeated before further discussion would be warranted.

results has been questioned by several people, both privately and in print [70.B; 78.F1]; however, it appears to me to be the most promising example, to date, of the phenomenon being described here, and an alternative demonstration, which we could accept with more confidence, would be most welcome.

# 6

# *A molecular dynamic approach to specific rate functions*

In the preceding chapter, we discovered that the fall-off behaviour for a large class of thermal unimolecular reactions could be reproduced more-or-less to perfection if the specific rate function was assumed to have the form given in equation (4.9). At first sight, this may seem to be a rather artificial form to choose but, in fact, as we will now see, when one attempts a state-to-state synthesis, a rather similar result ensues.

### 6.1 Every state has a different rate!

To begin with, let us consider one of the simplest possible chemical reactions to exhibit unimolecular reaction behaviour, the thermal decomposition of nitrous oxide to form a nitrogen molecule and an oxygen atom

$$N_2O \rightarrow N_2 + O$$

The algebraic and numerical details of the description I am about to give can be found in [79.Y3; 79.Y4]. The molecule is linear, having two rotational and four vibrational degrees of freedom; there are two independent stretching modes and a doubly degenerate bending mode [45.H]. In quantum mechanics, we can model the reaction process by calculating the matrix elements for transitions between given quantum states of the $N_2O$ molecule and any possible final states (i.e. rotation–vibration states of $N_2$ and relative translation between the $N_2$ molecule and the O atom). Let us make the simplifying assumption that these matrix elements are independent of (i) the overall rotational angular momentum of the reactant molecule, and (ii) the number of quanta residing in the pair of bending vibrational modes: this reduces enormously the number of transition matrix elements to be calculated.[1] I also introduce at this juncture a piece of philosophy upon which we will come to depend more and more: that when we consider the molecular motions which cause a chemical transformation to occur, a more natural description will result if we use something more akin to a local-mode or bond-mode formulation [81.S1] than the traditional normal-mode approximation to the internal motions. Hence, in this case, we model the interesting motions as a pair of coupled oscillators, one representing the elongation of the N–N bond, taken to be harmonic because of its very large dissociation energy, and the other representing the stretching of the N–O bond, taken to be a Morse oscillator.

There is, unfortunately, a slight complication about this particular process which does not arise in most unimolecular reactions: this reaction takes place by a spin-forbidden transition between two electronic states whose zeroth order potential curves cross each other. As a consequence, the transition matrix elements are made up of two basic parts. The first is the nuclear part which comes from the overlap of the initial vibrational wavefunction (representing the coupled oscillators having a specified pair of quantum numbers) with the final translation–vibration wavefunction (representing the nitrogen molecule in any energetically allowed vibrational state with the remainder of the initial energy in the relative translational recoil). The second part is the non-adiabatic electronic coupling term, whose net effect is to scale down the overall size of the matrix elements by between one and two orders of magnitude; apart from this, it does not seriously impede our attempt to formulate specific rate functions in state-to-state terms.

Figure 6.1 shows a selection of these matrix elements, expressed as rate constants for state-to-state processes

$$O–N–N \rightarrow O + N–N$$
$$\phantom{O–}v_{b1}\ \ v_{b2}\phantom{\rightarrow O + N}v_{u2}$$

where $v_{b1}$, $v_{b2}$, and $v_{u2}$ are the vibrational quantum numbers of the bonds as labelled: the horizontal axis is the total vibrational energy shared between the two oscillators, relative to the combined zero-point energy;

---

1 The reason for this is twofold, for not only is the number of initial states minimised, but the choice of rotational state for the product $N_2$ molecule drops out also. It has been pointed out, rightly, [80.L1] that this is a considerable assumption: reaction is considered to take place as a non-adiabatic transition between two electronic states of the $N_2O$ molecule, and although the ground state is linear, the other one is not; consequently, the bending motions should play a part in the reaction process. Thus, whilst one may regard the numerical results of the simpler treatment with some circumspection, it remains an ideal vehicle for illustrating the state-to-state synthesis of specific rate functions.

for this model calculation, there are 527 processes having rate constants $d_i$ greater than $10^5$ s$^{-1}$, only about 60 are shown here. Without entering into any attempt to rationalise the behaviour shown, I simply want to call attention to the principal feature of these results, that the calculated rate constants are highly oscillatory with a ceiling of about $10^{12}$ s$^{-1}$, which corresponds to a reactive event about every 30–50 vibrational periods.

## 6.2 The smoothing effect of transforming from $d_i$ to $k(E)$

Now the specific rate function $k(E)$ is not what is plotted in Figure 6.1: $k(E)$ is the mean rate constant to be applied to all reactant molecules having an internal energy in the range $E_r \leqslant E < E_{r+1}$, regardless of the

Fig. 6.1. State-to-state rate constants for the dissociation of $N_2O$; $v_{b1}$ and $v_{b2}$ are the quantum numbers for the N–O and N–N vibrations respectively, in the reactant molecule, and $v_{u2}$ is the quantum number for the vibration in the product $N_2$ molecule. The thermochemical threshold is at 13 488 cm$^{-1}$, but the microscopic rate constants are almost negligible below about 20 000 cm$^{-1}$. The symbols have the following meanings: crosses, $v_{b2} = v_{u2} = 0$; circles, $v_{b2} = 0$, $v_{u2} = 1$; triangles, $v_{b2} = 4$, $v_{u2} = 2$; asterisks, $v_{b2} = 4$, $v_{u2} = 3$.

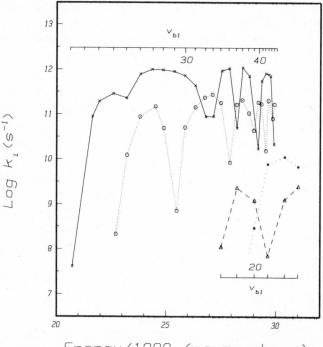

way in which that energy is distributed among the allowed degrees of freedom. To arrive at this quantity, recall that

$$k_\infty = \sum_i \tilde{n}_i d_i = \sum_i g_i d_i e^{-\varepsilon_i/kT} \tag{4.1a}$$

$$= \{Q(T)\}^{-1} \int_0^\infty \rho(E) k(E) e^{-E/RT} dE \tag{4.2}$$

and isolate the product $f(E) = g_i d_i \equiv \rho(E) k(E)$; now subdivide the range into grains of energy width $\Delta E$ with index $r$, and define the function

$$\phi_r = (\Delta E)^{-1} \sum_i g_{i,r} d_{i,r} \equiv (\Delta E)^{-1} \int_{E_r}^{E_{r+1}} f(E) dE \tag{6.1}$$

A grain width of $17.5 \text{ cm}^{-1}$ ($0.05 \text{ kcal mol}^{-1}$) is used for this development. In this way, by summing all of the decay rate constants weighted by the respective degeneracies of the reactant states, within the given grain, we arrive at the points shown as crosses in Figure 6.2: (remember that only about one-tenth of the necessary rate constants are shown in Figure 6.1) the oscillatory pattern persists. The quantity $\phi_r$ is now the mean rate constant for decay of molecules having an energy within the range $\Delta E$ at $E_r$, where $E_r$ is confined to be total vibrational energy shared by the two stretching modes.

Let us now go one stage further, and reinterpret $E_r$ to mean the total vibrational energy (in excess of the zero-point energy) of the molecule; in other words, $E_r$ may now also include the bending vibrational energy. In this model calculation, the bending frequency is almost exactly $600 \text{ cm}^{-1}$, and it is assumed to be simple harmonic. Now concentrate on one of these rate constants $\phi_r$, which is represented by a cross in Figure 6.2, at an energy $E_r$ – for example, the one of magnitude about $1.8 \times 10^{10} \text{ s}^{-1}$ $(\text{cm}^{-1})^{-1}$ at $E_r = 25\,000 \text{ cm}^{-1}$ with the bending modes unexcited. If the molecule is then considered to possess bending motions as well, there will be states of the molecule of *total* vibrational energy $25\,600$, $26\,200$, $26\,800$, $27\,400$, ... $\text{cm}^{-1}$, containing 1, 2, 3, 4, ... bending quanta in addition to the original $25\,000 \text{ cm}^{-1}$ of stretching motion. There being two independent bending modes, 1, 2, 3, 4, ... quanta of bending energy can occur in 2, 3, 4, 5, ... ways respectively, and so there is a contribution from this particular value of $\phi_r$ at each of these energies, with weight factors equal to the number of ways that the value of the energy can occur. What we now see in the reinterpreted $\phi_r$ function, represented by the triangles in Figure 6.2, is a steady increase because the higher the energy, the more ways there are in which we can find molecules having a certain amount of that energy fixed in such a way as to bring about reaction, with the remainder distributed at random throughout the other available internal degrees of freedom. Notice also that the scatter characteristic of the crosses is much reduced by this averaging process.

But, in fact, the variable $E$ which we use in the specific rate function $k(E)$ is the total energy of the molecule, including the rotational energy as well: consequently, we must repeat this process, convoluting the $\phi_r$ of Figure 6.2 with all the possible arrangements of the energy of rotation. This produces another stage of smoothing, and the results are shown as discrete points in Figure 6.3: this represents the final function $f(E)$ $\equiv \rho(E)k(E)$, given the approximations which were made in formulating the quantum mechanical model of the reaction process.

There is also another way of calculating $f(E)$ from the same fundamental data: we know from the calculated matrix elements all of the values of

Fig. 6.2. Grained reaction probability function $\phi_r$ for the dissociation of $N_2O$: crosses, vibrational motion only included; triangles, vibrational and bending motions included.

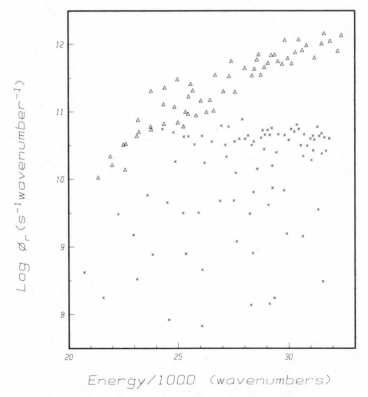

Energy/1000 (wavenumbers)

$d_i$, $g_i$, and $\varepsilon_i$ required to calculate $k_\infty$ from equation (4.1a), and by introducing a variation on the temperature $T$, we can plot a theoretical Arrhenius curve for this reaction. This plot of the theoretical values of $k_\infty$ is remarkably close to strict Arrhenius (provided the temperature is not extremely low [82.F1]), whence we can invoke equation (4.9) to yield

$$f(E) = \rho(E)k(E) = A_\infty \rho(E - E_\infty) \qquad (6.2)$$

which is plotted as the solid line in Figure 6.3; this is the gratifying result to which I have already alluded in Chapter 4.

### 6.3 An approximate synthesis for more-complicated molecules

Calculations like these on triatomic molecules [79.Y3; 80.L1] lie close to the limits of feasibility at the present time, and we cannot hope to

Fig. 6.3. Grained reaction probability function $f(E) \equiv \phi$, for the dissociation of $N_2O$; the discrete points represent a continuation of the calculation begun in Fig. 6.2, with all internal motions included, and the solid line is the same function derived from the inverse Laplace transform, i.e. equations (4.9) or (6.2).

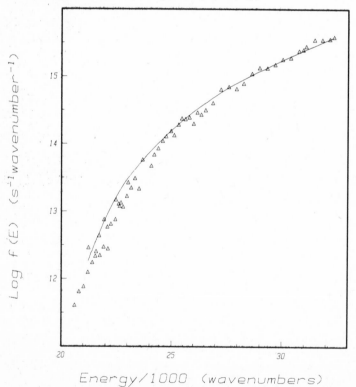

repeat them for molecules like methyl isocyanide or cyclopropane. The germ of the synthesis which I am about to describe was contained in a paper by Polanyi & Wigner in 1928 [28.P].[2] Let us assume that a simple unimolecular transformation takes place when sufficient vibrational energy finds itself concentrated in a single local or bond mode: for example, in a *cis–trans* isomerisation about a double bond, one would require that an amount of energy greater than the barrier height be located in the torsional vibration; since the torsional motion is quantised, this means that the molecule must be in a state of that torsional motion whose energy is greater than the height of the barrier to the torsional motion. Then, reinterpreting slightly the Polanyi–Wigner concept, we will accept that the transformation from reactant to product takes place within the vibrational period immediately following the instant at which the molecule was transferred into that state by the ever-present relaxation processes. Thus, the lifetime of any reactive molecular state is, on average, half of the period of the vibrational motion, whence its decay rate constant to products is

$$d_i = d = 2cv \qquad (6.3)$$

where $c$ is the velocity of light and $v$ is the frequency of the vibration in wavenumber units, regardless of the degree of excitation of the oscillator. This is a fundamental mechanical property of a classical oscillator, whether it be a harmonic or a Morse oscillator, with $v$ related to the potential constants in the usual ways, i.e.

$$2\pi cv = (k/\mu_n)^{\frac{1}{2}} \qquad \text{or} \qquad a(2D_e/\mu_n)^{\frac{1}{2}} \qquad (6.4)$$

respectively, where $\mu_n$ is the reduced mass of the nuclei, $k$ is the force constant of the harmonic oscillator, $D_e$ is the dissociation energy of the Morse oscillator, and $a$ is its anharmonicity constant [74.A].

A cursory inspection of Figure 6.1 might suggest that this is a gross oversimplification; however, the dominant contribution to the rate comes from those states of the $N_2O$ system in which all of the reaction energy is concentrated in the $N_2$–O vibration, and it would not do serious injustice to the data if it were to be assumed that every such state (apart from a few close to the reaction threshold) would decay with the same rate constant.

With this simplification, the quantity $f(E)$ for a given grain is just the

---

2 This paper is widely regarded as the origin of Slater's theory of unimolecular reactions. It seems to me ironic that Polanyi & Wigner were so close to a successful formulation of unimolecular reactions, but lacking the necessary theoretical apparatus at that time, they were forced to resort to a mechanical analogy one stage too soon: the end result, despite valiant efforts by Noel Slater, is a morass.

product of $d$ with the number of states within the grain which have more than the required critical energy in the specified oscillator; similarly, the specific rate constant for the grain, $k(E) = f(E)/\rho(E)$, is just $d$ multiplied by the fraction of such states within the grain. Figure 6.4 shows how the number of such states has to be counted. It is assumed that the $n$th quantum state of the oscillator is below the reaction threshold and that the $(n+1)$th state is above, and that we want to calculate $k(E)$ for an energy $E$ which lies between the energies of the $(n+4)$th and $(n+5)$th states of the oscillator. Let us suppose that the oscillator has $(n+4)$ quanta, and thus there is an amount of energy $(E - E_{n+4})$ distributed throughout the remainder of the degrees of freedom of the molecule: the number of such states is $\rho^r_{s-1}(E - E_{n+4})$, i.e. the density of states at the energy $(E - E_{n+4})$ for the rump of the molecule comprising $(s-1)$ of the original degrees of freedom with the designated oscillator $r$ removed. Likewise, there will be analogous contributions to $f(E)$ for the oscillator states $(n+3)$, $(n+2)$, and $(n+1)$, so that we may write the final expression for $k(E)$ itself as

$$k(E) = 2cg_r v_r \sum_i \rho^r_{s-1}(E - E_{n+i})/\rho_s(E) \qquad (6.5)$$

where $\rho_s(E)$ represents the total density of states of the reactant molecule

Fig. 6.4. Schematic representation of the disposition of states of the critical oscillator with respect to the threshold energy and with respect to the energy $E$ of the grain being considered.

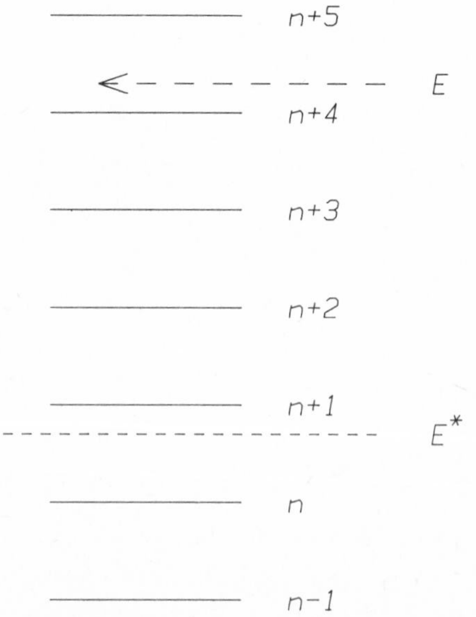

at energy $E$, and $g_r$ is the degeneracy of the oscillator (of frequency $v_r$) which is thought to lead to reaction when sufficiently excited.

### 6.4 Application to methyl isocyanide

There have been several theoretical studies of the potential energy surface for the transition between methyl isocyanide and methyl cyanide [68.V; 72.D2; 74.M], and it is clear that if one had to postulate a single vibrational mode which would, if sufficiently excited, lead to reaction, one would naturally choose the C–N–C bending mode $v_8$. This vibration has a frequency of $263 \, \text{cm}^{-1}$ and it is doubly degenerate; its degree of anharmonicity is not known, but numerical experiments (see Section 6.7 below) suggest that no great error is incurred by assuming that its energy levels are equally spaced. Figure 6.5 shows the comparison between the inverse Laplace transform estimate for the $k(E)$ function, and the result given by equation (6.5) with $v_8$ chosen as the reactive oscillator and

Fig. 6.5. Comparison of the synthetic specific rate function (dashed line) for the thermal isomerisation of methyl isocyanide with that derived from the inverse Laplace transform (solid line).

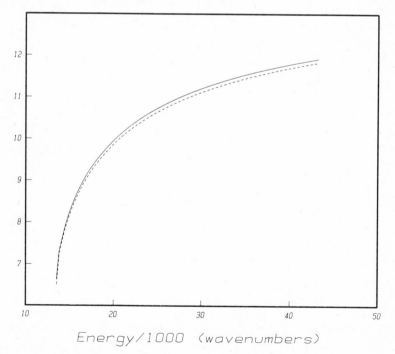

Energy/1000 (wavenumbers)

assumed to be harmonic; the agreement is remarkable.[3] It follows, naturally, that if the rate constant at infinite pressure is calculated by equations (4.1) or (4.2), it will also be in agreement with experiment and, in fact, we obtain $A_\infty = 3.6 \times 10^{13}$ s$^{-1}$ and $E_\infty = 38.3$ kcal mol$^{-1}$ to be compared with the observed values of $4.0 \times 10^{13}$ s$^{-1}$ and 38.4 kcal mol$^{-1}$ respectively; of course, agreement of the activation energies is to be expected, since the threshold energy (assumed to be $E_\infty$) must be known before equation (6.5) can be implemented. It is interesting to note that a comparison between equations (4.9) and (6.5) gives us an approximate identification of the nature of the experimentally observed quantity, the frequency factor

$$A_\infty \rho(E - E_\infty) \equiv 2cg_r v_r \sum_i{}' \rho^r_{s-1}(E - E_i) \tag{6.6}$$

where $\Sigma'_i$ denotes only those quantum states of the critical oscillator with energies greater than $E_\infty$. If the density of states term on the left of this equation is of a similar magnitude to the sum of the density of states terms on the right, then $A_\infty \cong 2cg_r v_r$, very much in line with the original ideas of Polanyi & Wigner [28.P] for the simplest reactions.

### 6.5 More-complex reaction co-ordinates

Many of the known unimolecular reactions have frequency factors in excess of $10^{15}$ s$^{-1}$ [72.R], and these clearly cannot be accommodated within the framework of equations (6.5) and (6.6): no transformation from the normal-mode to the local-mode description is going to yield an oscillation with a frequency much greater than $10^{14}$ s$^{-1}$, and so if the ideas described above are going to be applicable to these cases also, we must find an excuse for arguing that a greater fraction of the states within the grain are permitted to decay to products. The germ of the idea which might be used here is also contained in the original paper of Polanyi & Wigner [28.P]: they imagined that the reaction energy need not be confined to a single linear oscillator, and they showed that if this energy

---

3 It is important to realise how insensitive this degree of agreement is to the assumptions made about the spectral properties of the molecule. Arbitrary variations in the vibration frequencies, their anharmonicities, etc., cause only minute changes in the inverse Laplace transform $k(E)$ function for reasons already explained in Chapter 4, see Figure 1 of [80.P1]. The synthetic $k(E)$ function (equation (6.5)) is equally insensitive to these variations also, with the exception that it is very sensitive to the choice of the properties of the designated reactive oscillator, see, for example, Figures 3 and 4 of [80.P1]. These empirical observations put the *ab initio* calculation of a unimolecular reaction rate constant within our reach for a fairly simple molecule.

was shared among two or three degrees of freedom, the rate of reaction would be considerably enhanced.

The isomerisation of cyclopropane is a reaction to which we might apply this concept. Some calculations of the potential surface for this reaction have been made, and it was suggested that the form of the motion which leads to reaction must have both vibrational and torsional character [76.J]. In our tentative exploration of this idea, we assumed that the critical energy for reaction could be shared in any way between the ring stretch ($v_3 = 1188$ cm$^{-1}$) and *either* an asymmetric $CH_2$ stretch ($v_6 = 3103$ cm$^{-1}$) *or* $CH_2$ twist ($v_{13} = 1188$ cm$^{-1}$) (one or other of the degenerate pair in each case); the calculated values of $A_\infty$ from these assumptions are $1.6 \times 10^{15}$ and $1.4 \times 10^{15}$ s$^{-1}$ respectively, compared with an observed value of $1.9 \times 10^{15}$ s$^{-1}$. The requirement to concentrate all of the critical energy in a single oscillator gives rates at least an order of magnitude too small [80.P1].

It is clear that there is a lot of exploration to be done and a lot of thin ice to be trodden before this approach can be shaped into a form which has predictive capacity; nevertheless, the effort seems worthwhile.

## 6.6 Dissociation reactions

In laying his foundation of the theory of unimolecular reactions, Lindemann envisaged that the way in which a molecule became dissociated was by centrifugal forces becoming sufficiently large so as to overcome the binding forces between the nuclei [22.L]. After a long period of uncertainty, it has now become patently obvious that in diatomic dissociation reactions at high temperatures, rotational energy is equally as important as is vibrational energy in contributing to the reaction energy [75.P1; 79.Y1], and the same sort of considerations must be true for the dissociation of polyatomic molecules [76.P1]. We have therefore explored the idea that in a polyatomic dissociation reaction, the reaction energy may be considered to be shared in any way between a single bond vibration of Morse-like character and the pair of external rotational degrees of freedom which interact strongly with that vibration; rotation about the axis of the dissociating bond is considered to have only a weak effect upon the rate of the dissociation [80.J1]. We examined briefly four dissociation reactions, those of $CH_4$, $N_2O_5$, $C_2H_6$, and $C_2F_6$, and found in every case a frequency factor well in excess of $10^{15}$ s$^{-1}$; the computational method used, however, needs some refinement before definitive answers can be quoted.

This observation indicates that our earlier treatments of the decompositions of cyclobutane and of ethyl chloride [80.P1] are probably much too naive: both may be considered to be dissociation reactions, and so the overall rotation could be important in determining the rate; both reactions also involve curve crossings between different electronic states and it will take considerable theoretical effort to determine with confidence the relative magnitudes of these two opposing effects.

### 6.7 *Ab initio* unimolecular rate constants?

Our resurrection of the original approach of Polanyi & Wigner appears to me to provide, at least in simple cases, a direct molecular dynamic route to the calculation of unimolecular rate constants from theoretically constructed potential energy surfaces. Recalling the relative insensitivity of the specific rate function to energy-level spacings for motions other than the critical one,[3] it is clear that a very detailed knowledge of the potential surface is not required, at least initially. We would have to explore the potential surface in the region of the equilibrium nuclear configuration sufficiently well to be able to calculate the moments of inertia and normal modes of vibration of the molecule. This would give all the information needed to calculate $\rho(E)$ for the molecule, on the reasonable assumption that as the normal-mode description goes over to the local-mode description of the internal motions with increasing energy, the density of states given by either description is about the same. It is then necessary to explore a small segment of the potential surface lying between the equilibrium configuration and the products in enough detail to identify the nature of the local mode which leads to reaction, to calculate the energy levels of that mode and the normal frequency of its oscillation. It is apparent that the amount of computational labour required is not more than is already expended in contemporary reactive surface calculations.

As an example, our first move in this direction has been to solve for the restricted-rotor energy levels of the $CH_3-NC/CH_3-CN$ system given a theoretical potential for the torsional motion, and the calculated rate constant is very similar to that given in [80.P1]; in this trial [82.C] the remaining molecular frequencies were taken to be the experimental ones, but there is no reason to think that equally acceptable ones could not have been generated from an *ab initio* potential surface. We expect to extend our (so far) cursory examination of dissociation reactions in this manner also in the near future.

## 6.8 Experimental determination of $k(E)$

There have been several examples recently of experiments from which it is possible to deduce approximate values of the specific rate constant for reaction at certain fixed values of the energy $E$. If methyl isocyanide vapour is subjected to intense laser radiation at 7265 Å, corresponding to an excitation energy of 39.3 kcal mol$^{-1}$, then those molecules which absorb radiation are raised to an energy which is about 1 kcal mol$^{-1}$ above the thermal threshold for isomerisation.[4] In its most primitive form, the reaction sequence can be written as

$$\left. \begin{array}{ll} R + h\nu \rightarrow R^* & ; \text{rate} = [R] \times I_{abs} \\ R^* \rightarrow P & ; \text{rate} = [R^*] \times k(E) \\ M + R^* \rightarrow M + R & ; \text{rate} = [R^*] \times pr_i \end{array} \right\} \qquad (6.7)$$

Solution of the steady state equations gives the rate of formation of reaction products $R_P$ as

$$R_P = k(E) I_{abs} [R] / \{ k(E) + pr_i \} \qquad (6.8)$$

whence an assumption about the magnitude of the internal relaxation rate constant $r_i$, together with a knowledge of the absorbed intensity $I_{abs}$, yields a value of $k(E)$ for the energy in question. Molecules studied in this way include methyl and allyl isocyanides [77.R1; 79.R1; 79.R2], a series of cycloheptatrienes [78.H; 79.H], and tetramethyl dioxetane [81.C2].

A similar idea has been developed using chemical activation techniques: hot H atoms, formed from the photolysis of HBr or $H_2S$ at certain wavelengths, are allowed to react with 1-butene to form vibrationally hot $n$-butyl radicals; with the wavelengths used in these experiments [80.G1; 81.G], the $n$-butyl radicals, subject to the normal uncertainties in the assumed thermochemical data, were formed with excess internal energies of about 22, 28, 30, and 42 kcal mol$^{-1}$ above the reaction threshold.[5]

4 As I have noted before (Footnote 12 of Chapter 5), we do not really know the exact value of the energy at the threshold.

5 Notice, however, that the actual experimental data are not reduced by the use of an equation as simple as equation (6.8), since to do so would not, in general, give the true value of $k(E)$. This is because the final step in equation (6.7) assumes that any collision removes all excited molecules from the reactive region; this may be a reasonable approximation at energies close to the reaction threshold – or at any energy if there were really such a thing as a strong collision! In fact, as has been demonstrated in very many chemical activation

In all of these experiments, the derived values of $k(E)$ are reasonably close (usually well within an order of magnitude) to the theoretical function, calculated from the known (or assumed) rate law. There also exist, e.g. [82.B], elegant experiments in which the rate constants for decay of ionic species, as a function of the total energy, are measured in a time-of-flight mass-spectrometer, with a very similar degree of matching between theory and experiment.

experiments, the newly formed energetic molecules cascade down in a stepwise fashion, and so they are able to decay to products, with correspondingly reduced rate constants, for several collisional periods instead of only the one; it is this cascading effect which is responsible for the so-called 'low pressure turn-up', the apparently faster rate of decay at the lower pressures, in chemical activation studies [63.K3].

# 7

## Building in the randomisation processes

The general perception, all these years, of the course of a unimolecular reaction has been that the reactant molecule receives the requisite amount of energy by collision, but that it can only react to form products after a time delay during which it has to rearrange that energy and reassemble it into some more appropriate fashion.[1] This reorganisation of the internal energy has almost always been regarded as an intramolecular process [30.R], with a rate which is independent of all external conditions; except for this, the nature of the reorganisation process has been rather poorly defined. Broadly speaking, older treatments of the problem tend to visualise the molecule comprising two domains, one where the energy is collected from external sources by collision, and the other where it is needed to bring about the reaction, with a time delay for communication of internal energy between the two. On the other hand, many modern treatments use the language of radiationless processes: as such, they tend to be unnecessarily complicated *from the point of view of treating bulk unimolecular reactions*, and they also tend to neglect other possible causes which could have the same end result. The treatment given below is an elaboration of one which I have evolved [80.P2] and which I believe is helpful in clarifying the nature of the unimolecular reaction process.

### 7.1 A molecular model for treating randomisation in unimolecular reactions

Let us accept as a starting point that we can, in principle, write down all the rotation–vibration energy levels of the molecule in question. Just

1 Hinshelwood conjectured that there was also the possibility of a direct excitation process, which would leave the newly excited molecule in a state from which it must react: this would give rise to a small second order component to the reaction at high pressures, which Hinshelwood believed [58.B; 59.B], erroneously [68.S], that he had detected; it is now clear, in the light of the discussion given below, that such a second order component would be annihilated by the randomisation processes.

as in the case of a diatomic molecule, these are the non-adiabatic energy levels of the system. Let us also assume that if we could solve the Schrodinger equation sufficiently precisely, all of these levels would be distinct in energy, and their wavefunctions would all be mutually orthogonal; some of these levels could, of course, be degenerate because of molecular symmetry. We would then have the situation that in the absence of any external fields (i.e. with the molecule isolated in an infinitely large box whose walls were at a temperature of absolute zero), a molecule finding itself in a rotation–vibration level below the reaction threshold would remain in that level indefinitely. If the internal energy is not far below the threshold, then there will be very many closely spaced states (roughly $10^5$, $10^9$, $10^{15}$ states per wavenumber, respectively, in carbon dioxide, methyl isocyanide, and cyclobutane, for example), and we must now consider what external influences might cause such states to change into each other; all that is necessary is for the external influence to broaden a state sufficiently that its energy overlaps that of another state, and the transition may occur.

What might these external influences be? The first, and most obvious, is that even in the rigorous absence of an imposed electromagnetic field, there is a broadening of all molecular energy levels due to spontaneous emission, and this broadening by far exceeds the spacings between our hypothetical non-adiabatic molecular levels. In principle, therefore, the conversion of one state into an adjacent one can be described by a spontaneous emission followed by reabsorption of the same quantum [54.H1]. Likewise, the ever-present black body radiation will make an important contribution: such effects have only recently been documented, but in the case of highly excited states of the sodium atom, where the levels are several wavenumbers apart, ambient black body radiation at 300 K is as effective as is 10 mTorr of argon in causing transitions between states [79.B1; 80.C]. By their spectroscopic nature, both of these mechanisms will be subject to angular momentum restrictions through the optical selection rules and so, on the face of it, they would not seem to be the main cause of the wholesale reorganisation of internal motions, which appears to be necessary in strong collision systems. On the other hand, both processes would be independent of pressure and would give the appearance of being intramolecular in character.[2,3]

2 We have not returned, full circle, however, to the old radiation hypothesis: the contrary arguments of Langmuir [20.L] still stand, but they can be put on a much firmer quantitative footing these days when photon cross-sections or extinction coefficients are known for many infra red absorption lines. The rate of black body photochemical

A third mechanism is a collisional one, which has hitherto not been recognised in any general way as an integral part of the unimolecular reaction process; quite a few experiments now exist which indicate a collisional component to the so-called intersystem crossing between electronic states of complex molecules, but suggestions of a similar mechanism for the randomisation processes in unimolecular reactions have been rather tentative or conjectural [66.P; 75.N; 77.R1; 79.O; 80.W]. If we consider the Lennard–Jones interactions between pairs of common organic molecules [54.H2], we find that they exceed the spacings between the molecular energy levels even at very considerable distances: typical values for the interaction potential as a function of internuclear separation are shown in Table 7.1, together with the approximate values of the pressure corresponding to these separations.

It is obvious that if the intermolecular potential exceeds the mean spacing between the molecular levels, we cannot rule out the scrambling of the molecular states by what is, in effect, pressure broadening; however, since

activation is far too slow [27.L], and the only conditions under which it might compete with collisional excitation would be at interstellar molecular densities [79.D]. For the reader who is interested in this problem, this remark may be used as the basis for a small research exercise.

3 This fleeting discussion of radiationless transitions has a rather different flavour from that which is usually offered; I will cite only two of the many reviews available in this very rapidly progressing field, as convenient entry points for the student of chemical kinetics [78.F2; 82.P1]. It is common to begin by formulating a set of Born–Oppenheimer states since, in principle, the exact non-adiabatic states can be represented as a sum over Born–Oppenheimer states: however, the wavefunction of the Born–Oppenheimer state is a very poor approximation to that of the corresponding exact state [80.M]; conversely, one needs an enormous number of zero order wavefunctions to describe the wavefunction of an exact state [80.F2]. For example, in the simplest possible case, that of the hydrogen molecule–ion, the difference in energy between an adiabatic and non-adiabatic representation of the same state is of the order of 3 cm$^{-1}$ in the reactive region [77.K; 79.P1]: our problem here is to describe the broadening of two states initially 10$^{-6}$ cm$^{-1}$ (say) apart, when the wavefunctions we are using can only represent the energies of those states to within, at best, about 10 cm$^{-1}$; there is no guarantee that the computed result will bear a strong relationship with the desired one.

The student of chemical kinetics wishing to explore these matters will soon find his or her way to the classic paper of O. K. Rice in 1933 [33.R] via the reformulation of Fano [61.F2] and further development by Bixon & Jortner [68.B1]. Stepping stones in the treatment of randomisation in unimolecular reactions include those of Gill & Laidler [59.G], Slater [59.S2; 67.S1], Mies & Krauss [66.M], Solc [67.S2], Gelbart, S. A. Rice & Freed [70.G1], O. K. Rice [71.R1], Bunker & Hase [73.B2], but the approaches represented are quite varied; a review of some of these ideas can be found in [79.O].

Of fundamental importance for a proper understanding of unimolecular reactions is the difference between small molecule and large molecule behaviour [81.S3] in the sense that these terms are used in the discussion of intramolecular relaxation processes: this classification into small and large is closely parallel to our division into weak and strong.

Table 7.1. *Mean internuclear separation and Lennard–Jones interaction potential as a function of pressure*

| $p[\text{Torr}]$ | $\bar{r}[\text{cm}]$ | $V(\bar{r})[\text{cm}^{-1}]$ |
|---|---|---|
| $10^3$ | $3 \times 10^{-7}$ | $3 \times 10^{-1}$ |
| $1$ | $3 \times 10^{-6}$ | $3 \times 10^{-7}$ |
| $10^{-3}$ | $3 \times 10^{-5}$ | $3 \times 10^{-13}$ |

Table 7.2. *Approximate pressure limits for collisionless regimes as a function of molecular complexity*

| Molecule | $\rho(E_\infty)$ (cm) | $p(\rho)$ (Torr) |
|---|---|---|
| $N_2O$ | $4 \times 10^4$ | $10$ |
| $CO_2$ | $4 \times 10^5$ | $3$ |
| $CH_3NC$ | $2 \times 10^8$ | $1 \times 10^{-1}$ |
| $CD_3NC$ | $1 \times 10^9$ | $6 \times 10^{-2}$ |
| $C_2H_5NC$ | $1 \times 10^{12}$ | $2 \times 10^{-3}$ |
| $C_2H_5Cl$ | $4 \times 10^{12}$ | $1 \times 10^{-3}$ |
| cyclo–$C_3H_6$ | $4 \times 10^{12}$ | $1 \times 10^{-3}$ |
| $C_2H_6$ | $7 \times 10^{12}$ | $8 \times 10^{-4}$ |
| cyclo–$C_4H_8$ | $2 \times 10^{15}$ | $5 \times 10^{-5}$ |
| $C_2F_6$ | $4 \times 10^{21}$ | $3 \times 10^{-8}$ |

$\rho(E_\infty)$ is the density of rotation–vibration states near the thermal threshold and $p(\rho)$ is the approximate pressure for which a typical Lennard–Jones interaction potential becomes commensurate with the mean spacing between the states: for this calculation, I have taken $\varepsilon/k = 500$ K and $\sigma = 5$ Å, in the notation of [54.H2].

the density of molecular states varies strongly with increasing energy, it is not possible to define a truly collisionless regime for a molecule unless the energy of interest is also specified. We are concerned here with uni-molecular reaction from molecular states with $RT$ or so above threshold and hence, *for this application*, we should use the densities of states for those particular energies. Table 7.2 shows, for a selection of interesting molecules, the approximate density of states near threshold and the pressure for which the mean intermolecular potential is the same as the mean energy spacing between the states; consequently, above these pressures, collisional randomisation must be considered as a serious possibility. It should also be pointed out that if these randomising collisions take place at distances of a few molecular diameters, or more, they will have enormous impact parameters; in such circumstances, even minute interchanges of energy could give rise to quite significant changes in angular momentum, so that many rotational states of the molecule at the energy in question may be coupled by this mechanism.

An alternative way of trying to estimate the onset of collisional randomisation is to use known collisional broadening factors: for example, in methane, the magnitude of the line broadening is $0.086 \, cm^{-1} \, atm^{-1}$ for one of the observed transitions [79.G]; this translates into just over $10^{-4} \, cm^{-1} \, Torr^{-1}$, not inconsistent with the earlier entries in Table 7.2.

## 7.2 Randomisation and reaction

We need only extend this model slightly to take account of the reactive states above threshold: these states decay to products within one characteristic vibrational period, as we argued in the preceding chapter. Thus, a molecule finding itself in such a state will be transformed into products (unless it is transformed into some other nearby state by one or other of the randomising mechanisms before the vibrational period is over). On the other hand, a molecule finding itself in any other state will not decay, but will be subjected to a continual cycling among all the states of approximately the same energy by the randomising mechanisms; most of these states are of the unreactive kind, as we saw in the previous chapter, but occasionally it will be cycled into a reactive state, and decay. We do not yet know the extent of the range of energy over which these stirring mechanisms can operate, but a convenient way to proceed (to begin with, at least) is to imagine that the domain over which the randomising processes are effective coincides with the grain which we have already chosen for computational purposes; conversely, the randomisation processes are assumed not to operate across the grain boundaries, which is an obvious approximation. In principle, we can make the energy width of the grains as small as we wish, but in practice the grains will always be far too large and, if we use too few of them, the calculated result may become sensitive to the grain size.

I want to proceed initially in an intuitive manner in which collisional transfer of molecules between grains takes place only by the usual strong collision processes with mean rate $\mu$. Stirring within the grains takes place by three distinct mechanisms, the standard collisional one with mean rate $\mu$, together with the first order and second order randomising mechanisms which I have already postulated, with mean rates $\mu_1$ and $\mu_2$ respectively; I have also argued that there is no danger in assuming that individually these processes may be regarded as pure exponential relaxations, so that we may continue to draw on strong collision formulae already developed in the earlier chapters [80.P2]. We can now postulate a mean relaxation rate for the grain as

$$\mu_g = \mu + \mu_r = \mu + \mu_1 + \mu_2 = pr_i + \mu_1 + pr_2 \tag{7.1}$$

where $r_i$ is, as before, the rate constant for relaxation of rotation–vibration energy in the molecule, $\mu_1$ and $r_2$ are the rate constants for first and second order randomisation processes respectively, and $p$ is the pressure. Moreover, since the internal randomisation rates are always very much faster than the internal relaxation rates, we can make an effective separation of the two types of process by writing

$$\mu_g \simeq \mu_r = \mu_1 + \mu_2 = \mu_1 + pr_2 \tag{7.2}$$

There is no reason why the randomisation rate $\mu_r$ should not be regarded as being different for each grain, but in all the numerical experiments I have conducted so far, I have taken it to be constant for a given molecule; also, it is clearly an approximation to regard $\mu_g$ in equation (7.1), or $\mu_r$ in equation (7.2), as strict exponential decay rates, but the error in so doing is, I hope, small.

We start with this picture: the molecule is left in grain $r$ as the result of a normal collision and, as the colliding partner recedes, it is being stirred continuously among all the states in the grain; if the partner were to recede to infinity, then the molecule would be left in one of our hypothetical non-adiabatic states. Once the collision is over (and remember that we are interested in a reaction process which takes place in the relatively long periods between collisions), cycling between the states within the grain occurs at the rate $\mu_r$, and decay from a subset of those states may also take place, with a rate constant $d$ as defined in equation (6.3), the same for each of the reactive states within the grain. Consequently, we can regard the decay of the grain to products as its own isolated unimolecular reaction problem, and we can use equation (5.22) to find the effective rate constant, viz.

$$k(E) \equiv d_r = \phi' d\mu_r/(\mu_r + d) \tag{7.3}$$

where $\phi'$ is the fraction of reactive states within the grain. Equation (7.3) is of strict Lindemann form (in $\mu_r$), because $d$ is constant, and in the limit as $\mu_r \to \infty$, $k(E) \to \phi'd$ which is, in fact, exactly the same as the expression, equation (6.5), we derived from a synthetic approach in the last chapter. We have arrived at a very interesting conclusion: if the randomisation rate is very large ($\mu_r > 100d$), then the grain will exhibit a value of $k(E)$ $= \phi'd$ which is usually called the statistical limit; if, on the other hand, $\mu_r$ is comparable with or less than $d$, then the effective value of $k(E)$ will be below the statistical limit. As a result, if the randomisation processes are rate limiting, the values of the overall reaction rate constant calculated by equation (5.14) will be perturbed; the precise nature of this perturbation will depend upon whether or not $\mu_r$ is pressure dependent, and this is illustrated in Figures 7.1 and 7.2. Figure 7.1 shows what happens in a model calculation on the isomerisation of methyl isocyanide if $\mu_r$ is given one of five fixed values, $\infty$, $10^{13}$, $10^{12}$, $10^{11}$, and $10^{10}$ s$^{-1}$, respectively, from the top of the diagram downwards; the high pressure limiting rate is depressed because $d_r < \phi'd$, but all curves merge towards the left because $k_{uni}$ is independent of $d_r$ at the low pressure limit. Figure 7.2 shows the

Fig. 7.1. Fall-off curves for a model calculation on the thermal isomerisation of methyl isocyanide considering only first order randomisation processes. In descending order, the curves correspond to values of $\mu_1 = \infty$, $10^{13}$, $10^{12}$, $10^{11}$ and $10^{10}$ s$^{-1}$ respectively.

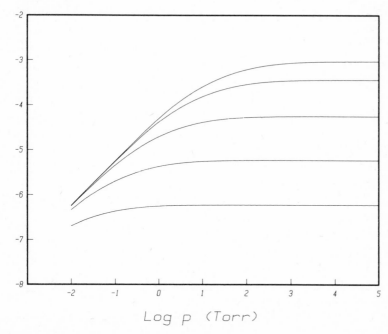

Log $p$ (Torr)

analogous calculation for $\mu_r = pr_2$, with (again from top to bottom) $r_2$ = $\infty$, $10^{13}$, $10^{12}$, $10^{11}$, and $10^{10}$ Torr$^{-1}$ s$^{-1}$ respectively; in this case, all high pressure limiting rates coincide because $p$ can always be chosen so that $pr_2 > d$, but they fall away at the low pressure limit because $d_r$ itself now possesses an additional pressure dependence which does not drop out [80.P2].

It is quite clear from an inspection of Figures 7.1 and 7.2 that the kind of agreement found between theory and experiment in Chapter 5 could not have occurred unless $\mu_r$ was effectively infinite in these strong collision systems, because as soon as the randomisation processes become rate determining, there are severe departures from the simple strong collision, i.e. $\mu_r = \infty$, behaviour. We must then ask whether there are any known experimental results from which we can demonstrate a departure from infinitely rapid randomisation, and it appears that the well-studied thermal isomerisation of methyl isocyanide may be just such an example.

Fig. 7.2. Fall-off curves for a model calculation on the thermal isomerisation of methyl isocyanide considering only second order randomisation processes. In descending order, the curves correspond to values of $r_2 = \infty$, $10^{13}$, $10^{12}$, $10^{11}$ and $10^{10}$ Torr$^{-1}$ s$^{-1}$ respectively.

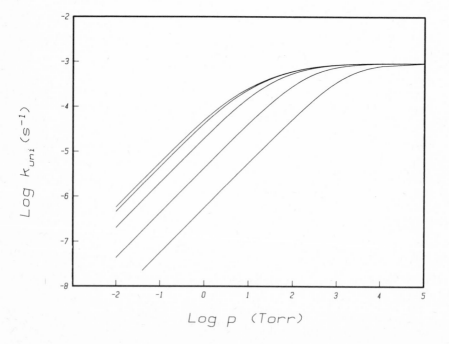

## 7.3 Limited randomisation in the thermal isomerisation of methyl isocyanide

The thermal isomerisation of methyl isocyanide is now the most comprehensively studied unimolecular reaction known. However, whenever comparisons have been made between theoretical representations of the fall-off and the actual experimental results, there were always small but real and tantalising discrepancies [62.S; 72.F1; 72.F2; 74.C; 78.Y2]. Figure 7.3 shows the experimental results of Schneider & Rabinovitch for the temperature of 230.4°C [62.S] compared with the standard strong collision fall-off (dashed line) having the same high and low pressure limiting rates; the discrepancy certainly exceeds the statistical uncertainty of the measurements. The solid line which passes adequately through the experimental points is calculated by assuming a finite value of $\mu_r$, equation (7.2), with $\mu_1 = 4.0 \times 10^{12} \, s^{-1}$ and $r_2 = 1.25 \times 10^{12} \, \text{Torr}^{-1} s^{-1}$. This gives rise to an effective $k(E)$ value for each pressure via equation (7.3), and substitution of this into equation (5.14) gives the desired result; the values of $\mu_1$ and $\mu_2$ were chosen simply to give the best fit to the data, without any preconceived notions as to what they ought to be. Measurements also exist of the fall-off in rate for the fully deuteriated methyl isocyanide [63.S1], and there is a similar discrepancy with the strong collision calculation, as shown in Figure 7.4: in this case, the best fit to the data is obtained with $\mu_1 = 5.0 \times 10^{12} \, s^{-1}$ and $r_2 = 3.0 \times 10^{12} \, \text{Torr}^{-1} s^{-1}$; those constants which give the best fit for the normal molecule in Figure 7.3 do not give an acceptable representation of the fall-off for the deuteriated molecule [80.P2].

Schneider & Rabinovitch [63.S1] also performed another series of experiments in which they measured the relative rates of isomerisation of the two isotopic species of methyl isocyanide: Figure 7.5 demonstrates the remarkable internal consistency of their data and shows that one does not expect to find the simple S-shaped dependence on pressure which is predicted by strong collision theory.

These values of the first and second order randomisation rate constants do not seem unreasonable. The values for $\mu_1$ are about the same for both molecules, and are close to the generally accepted estimate of around $10^{12} \, s^{-1}$ for intramolecular randomisation processes [82.T]. No previous estimates have been made for second order randomisation rate constants, but it seems plausible that they might increase with increasing state density, as we find here: the density of states near threshold is about six or seven times greater in the deuteriated as opposed to the normal methyl

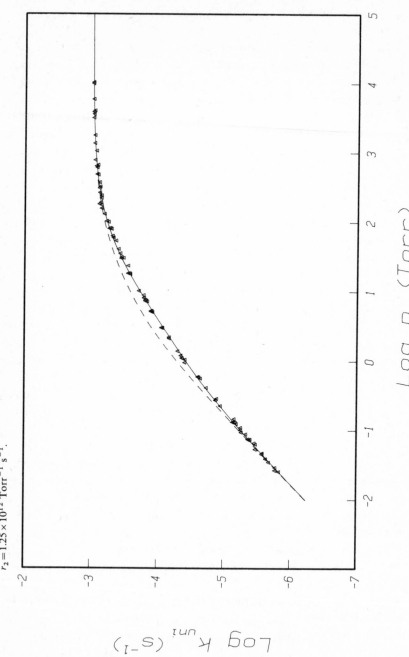

Fig. 7.3. Fall-off curves for the thermal isomerisation of methyl isocyanide at 230.4 °C. The points represent the experimental results of Schneider & Rabinovitch [62.S]. The dashed line is the standard strong collision fall-off calculation. The solid line includes both first order and second order randomisation effects, with $r_i = 1.20 \times 10^6$ Torr$^{-1}$ s$^{-1}$, $\mu_1 = 4.0 \times 10^{12}$ s$^{-1}$, and $r_2 = 1.25 \times 10^{12}$ Torr$^{-1}$ s$^{-1}$.

direct fall-off measurements of Schneider & Rabinovitch [63.S1] for this reaction. The dashed line is the standard strong collision fall-off calculation. The solid line includes both first order and second order randomisation effects, with $r_i = -1.30 \times 10^6 \text{ Torr}^{-1} \text{ s}^{-1}$, $\mu_1 = 5.0 \times 10^{12} \text{ s}^{-1}$ and $r_2 = 3.0 \times 10^{12} \text{ Torr}^{-1} \text{ s}^{-1}$. The dotted curve shows the result of using the same randomisation rate constants as were used in the preceding diagram.

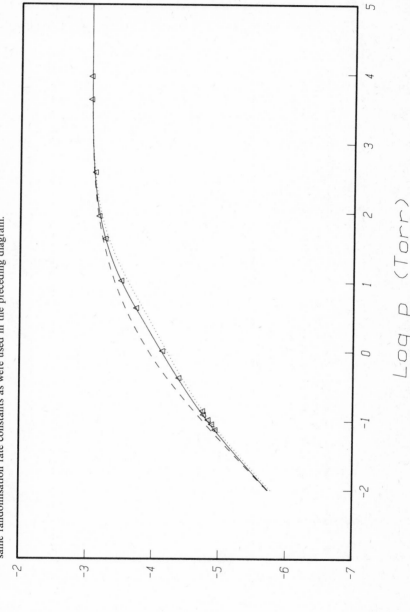

Log $P$ (Torr)

Log $k_{uni}$ (s$^{-1}$)

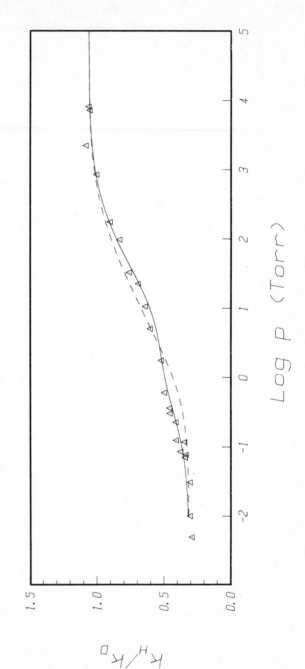

Fig. 7.5. Kinetic isotope effect $k_H/k_D$ for $CH_3NC$ and $CD_3NC$ thermal isomerisations as a function of pressure at 230.4°C. The points represent two sets of ratio measurements made by Schneider & Rabinovitch [63.S1]. The dashed line uses the standard strong collision fall-off calculation for each molecule, whereas the solid line allows for reactant state randomisation with the rate constants used in Figures 7.3 and 7.4 respectively.

isocyanide, and $r_2$ is some two to three times as fast. Unfortunately, there are no other reactions for which sufficiently comprehensive experimental results exist, and so it is not possible to extend these tests to molecules of different sizes at the present time.[4]

The dotted line in Figure 7.4 reveals that, for a given form of $\mu_r$, the magnitude of the anomaly in the fall-off curve is not fixed: if we could shift the fall-off curve to lower pressures for a given $\mu_r$, the 'feature' would become more prominent and, conversely, if we could shift it to higher pressures, it would disappear. Unfortunately, it is not possible in practice to change $r_i$ without also affecting the behaviour of $\mu_r$, and the only experiment in existence which throws any light on the problem is analysed in Figure 7.6. Wang & Rabinovitch have measured an almost complete fall-off curve for the methyl isocyanide reaction, with helium as the pressurising gas [74.W2; 75.W]; since helium is about one-sixth as efficient in sustaining the unimolecular rate as is methyl isocyanide itself [70.C], the fall-off curve is shifted to pressures which are about six times the normal values. The value of $r_i$, which is needed to position the curve properly, is (to within the accuracy of the calculation) exactly the value used in Figure 7.3 multiplied by the relative efficiency for helium, measured in [70.C]. The value of $\mu_1$ (which also includes a fixed second order contribution from the 0.01 Torr of reactant which is always present) is also indistinguishable from that used in Figure 7.3. On the other hand, the value of $r_2$ is a little more than a factor of 10 smaller than that for pure methyl isocyanide, and this causes the fall-off curves to be further apart horizontally at high pressures than at low pressures; this effect was accounted for previously in terms of the idea that the collision efficiency of helium declined as the high pressure limit was approached because the average energy of the reacting molecules is greater [77.T1]. A third, and serious possibility (see Section 9.5) is the failure of the mixture rule for internal relaxation at very high dilutions of methyl isocyanide in helium. More experiments of this nature are needed to decide which of these three interpretations is to be preferred.

## 7.4 Randomisation among states of the product

In the argument so far, we have concentrated upon randomisation effects in the reactant molecules, almost as though randomisation within

4 The fall-off curve for the thermal decomposition of 1,1-difluorocyclopropane [79.C] appears to exhibit a rather more distinct feature than we see in Figures 7.3, 7.4, and 7.6; however, in view of the general difficulties in handling fluorocarbons at elevated temperatures, it may be premature to try to interpret these experiments in a similar way.

the product molecules was irrelevant: this is not so, but we have to distinguish carefully between two quite different cases, fragmentation reactions and isomerisation reactions.

In the case of a fragmentation reaction, we imagine the separation to take place by the extension of the molecule along a particular bond mode, for example the C–C stretch in the dissociation of ethane into two methyl radicals; more-complicated motions may, however, be necessary to account for other fragmentation reactions such as the formation of two ethylene molecules from cyclobutane [80.P1]. Suppose that the molecule is sufficiently excited that it has enough energy to dissociate and, moreover, this energy is concentrated in the appropriate bond mode; suppose also that the two fragments are moving away from each other. There are two possibilities. As the result of a randomising influence, such as we have already described, the molecule may be cycled into another

Fig. 7.6. Fall-off curves for the thermal isomerisation of methyl isocyanide at 0.01 Torr in the presence of helium, at 245°C. The dashed line represents the strong collision fall-off curve, whereas the solid line is calculated with $r_i = 2.2 \times 10^5 \, \text{Torr}^{-1} \, \text{s}^{-1}$, $\mu_1 = 5.0 \times 10^{12} \, \text{s}^{-1}$, and $r_2 = 1.0 \times 10^{11} \, \text{Torr}^{-1} \, \text{s}^{-1}$. The points are the measurements of Wang & Rabinovitch [74.W2; 75.W], but they do not extrapolate well towards the recommended value of $k_\infty$ [62.S], and it has been assumed in constructing this diagram that the temperature was 243°C.

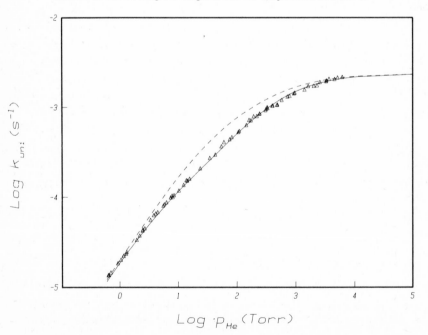

state in which it is unreactive, but this could only happen if the interfragment distance was sufficiently short for the interfragment potential to be attractive; due allowance for this eventuality is already included in equation (7.3). Alternatively, if the oscillation has developed to the extent that the two fragments are now on the repulsive side of the appropriate rotational barrier and accelerating away from each other, dissociation becomes inevitable: the weak randomising influences, as we have imagined them, would be incapable of reversing the relative motion of the two massive particles once they are receding in this manner; their relative motion can only be reversed by a collision with another body, which would constitute, in fact, the back-reaction.

Randomisation of the products does not cease, however, once the fragmentation becomes irrevocable: there will, in general, be a significant repulsion between the two fragments extending over a distance of several molecular diameters, and this will cause a mutual stirring of the two fragments among their respective states without necessarily requiring any significant interchange of energy between them. This prolonged randomisation within the products cannot have any effect upon the overall rate of the reaction.

The situation is quite different in an isomerisation reaction, where randomisation among the product states is an integral part of the reaction process. Figure 7.7 depicts the variation of the potential energy of the $CH_3$–(NC) system as a function of the C–(NC) angle: the energy levels in the left well represent the torsional levels of the methyl isocyanide molecule, and those in the right well, those of methyl cyanide. Immediately above the barrier, the levels represent a restricted rotation of the (NC) group with respect to the methyl radical, and as the energy becomes higher, this motion quickly transforms into an essentially free rotation. Let us examine any one of these restricted rotor states: with the exception of one or two states just above the barrier, the molecule spends roughly half its time in spatial configurations which have all the attributes of a methyl isocyanide molecule, and the remainder of the time in which the spatial configurations clearly correspond to those of the methyl cyanide molecule. Our zeroth order picture of the system is that it will remain in this state until it is caused to degrade into another state of the whole molecule at about the same energy, by one of the randomising influences, and then, depending upon what the actual nuclear configuration was at the moment of this occurrence, the system either becomes a product cyanide molecule or else it reverts to the original reactant isocyanide molecule. At this point, the notion that the randomisation rate constant $\mu_r$ is strongly dependent upon the density of states is important: if the hindered rotor state stood an equal chance of being randomised into either molecule, simply because it spent roughly half its time in either nuclear configuration, then $k(E)$ and the overall rate would be just half of what was calculated in Sections 6.4 and 6.5. However, at the same value of the total energy just above threshold, methyl cyanide has almost two orders of magnitude more states than does methyl isocyanide, simply because the ground rotation–vibration state of the cyanide is much lower down; assuming that this gives rise to a greater

Fig. 7.7. Disposition of the torsional states of the CH$_3$–(NC) system. The potential energy for this torsional motion is shown schematically as a function of the torsional angle; the well centred at zero represents the configurations of the CH$_3$NC molecule (reactant), and the well centred at $\pi$ represents the configurations of the CH$_3$CN molecule (product). A selection of the energy levels in the region of the thermal threshold is shown, with the positions of the energy levels depicted by horizontal lines. The significance of the dashed lines is as follows: with the torsional potential used in this calculation [82.C], the topmost level in the reactant well is a tunnelling state, with about a 3% chance of being found on the product side; also, in the first two states above the barrier, the rotational motion is significantly restricted, with the first one being reactant 23% of the time, and the second one being product 37% of the time. All of the higher levels possess the same rotational quality in that they are about 43% reactant, 57% product; more time is spent in the product configuration because the right-hand well is considerably deeper. The shading on this diagram is intended to remind the reader that there are very many other states of these two molecules at these particular energies, with the stronger shading on the right-hand side representing the fact that CH$_3$CN has a greater density of states than does CH$_3$NC at the same total energy.

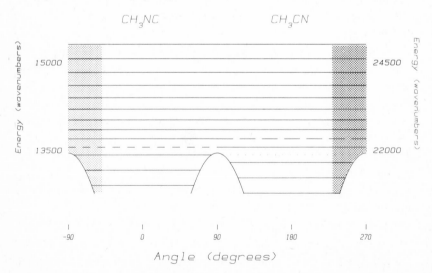

rate of randomisation out of the rotor state into the cyanide configurations than into the isocyanide configurations, then most of the molecules finding themselves in this state will end up as cyanide, and the reaction will develop the full rate that we predicted above.

We conclude that randomisation of the product states is a vital part of any isomerisation reaction. This being the case, then we may also conclude that at such time in the future when we can calculate rate constants with sufficient confidence, the rates of thermoneutral isomerisations will need to be reduced by a factor of about two compared with the rates of strongly exothermic isomerisations.

### 7.5 Separability of the relaxation and the randomisation processes

This treatment of the interplay of relaxation and randomisation in thermal unimolecular reactions has assumed that the two processes are completely separable, on the rather plausible grounds that the randomisation rates are very much faster than the relaxation rate. The conditions for such separability are, in fact, known. We can write the full master equation for reaction as

$$d\boldsymbol{\eta}(t)/dt = [Q + Q' - D]\boldsymbol{\eta}(t) \qquad (7.4)$$

where $Q$ and $D$ have their usual meanings, and $Q'$ is a matrix containing the rate constants for the randomisation processes; as far as I am aware, D. J. Wilson was the first person to try to formulate the problem in this fashion [60.W]. In the picture of the reaction process that we have developed above, $Q$ is a full matrix connecting all the levels in the system, but $Q'$ consists of a series of small blocks, each connecting only levels within a very small energy domain, i.e. within the grain itself; thus $Q'$ is of block-diagonal form. Later, Nordholm showed [76.N] that the dynamics of the internal (randomisation) processes would be separable from those of the external (relaxation) processes if, and only if, the matrix $Q$ were to commute with the matrix $[Q' - D]$, which is, in general, quite impossible. However, $Q$ and $Q'$ commute: the zero eigenvector of $Q$ is $S_0$; likewise, because the randomisation processes will establish a Boltzmann distribution within the domain of the grain, the zero eigenvector of each block within $Q'$ is the appropriate segment of $S_0$, and the concatenation of all of these segments is $S_0$ itself. Thus, the matrices $Q$ and $Q'$ share a common eigenvector, and therefore they commute. Moreover, as we saw in Chapter 3, if the elements of $D$ are small compared with those of $Q'$, then the eigenvectors of $[Q' - D]$ will differ only slightly from those of $Q'$ itself, and so the $Q$ will commute with $[Q'-D]$ in the limit as $\mu_r \to \infty$; thus, our intuitive separation of the internal and the external processes in the methyl isocyanide reaction, on the grounds that the internal processes are very much faster, is entirely reasonable.

### 7.6 An exactly solvable case

If the decay towards equilibrium caused by the randomisation processes can be assumed to be exponential (see Section 7.2 above), then the solution to equation (7.4) can be expressed in closed form. Let $Q$ be our usual strong collision relaxation matrix with relaxation rate $\mu$, and let the subblocks of $Q'$ each be of strong collision form, with relaxation rates $\mu_r$, where $r$ is the index denoting the ordering of the grains. The reaction matrix of equation (7.4) is symmetrised by the usual procedure

$$-E^{-\frac{1}{2}}[Q+Q'-D]E^{\frac{1}{2}} = [\mu(1-p_0)+V+D]$$
$$= [\mu(1-p_0)+\Sigma_r(V_r+D_r)] \qquad (7.5)$$

where $V_r$ is the appropriate symmetrised form of the strong collision relaxation matrix representing the randomisation process for grain $r$, and $D_r$ is the segment of $D$ lying within the domain of $r$; remember that in this formulation, the elements of $D$ are either zero or $d$ only. The following are the key points in the solution, which is given in detail elsewhere [81.P2]. Each block $(V_r+D_r)$ has four eigenvalues

$$\kappa_{r0} < \kappa_{r2} = \mu_r < \kappa_{r1} < \kappa_{r3} = \mu_r + d$$

$\kappa_{r2}$ and $\kappa_{r3}$ have multiple degeneracies (as already noted in Section 5.3), whereas $\kappa_{r0}$ and $\kappa_{r1}$ are simple (degeneracy of one) with values given by

$$\kappa_{r0,1} = \frac{\mu_r+d}{2} \mp \left[\frac{(\mu_r+d)^2}{4} - \frac{\mu_r\beta_{r1}d}{\beta_r}\right]^{\frac{1}{2}} \qquad (7.6)$$

where $\beta_r$, as before, is the equilibrium population of the molecules in grain $r$, and $\beta_{r1}$ is the equilibrium population of the reactive states within the grain. Notice that the term $\beta_{r1}d/\beta_r$ in the square root is the statistical limit for $k(E) \equiv d_r$ derived in the preceding chapter. Also, under conditions when $\beta_r \gg \beta_{r1}$, the expression for $\kappa_{r0}$ given by equation (7.6) reduces to the standard Lindemann form (7.3). An iterative procedure can then be developed [81.P2] which will converge to the exact eigenvalue $\gamma_0$, i.e. the rate constant, but under the usual conditions where $\mu \gg \gamma_0$, we can obtain an upper bound, which is an excellent approximation to the true eigenvalue, in the form

$$\gamma_{ap} = X/(1-\mu^{-1}X) \qquad (5.17a)$$

$$X = \sum_r' \sum_{j=0}^{1} \frac{\left[\beta_{r0} + \beta_{r1}\left(1 + \dfrac{d}{\mu_r - \kappa_{rj}}\right)^{-1}\right]^2}{\left[\beta_{r0} + \beta_{r1}\left(1 + \dfrac{d}{\mu_r - \kappa_{rj}}\right)^{-2}\right]} \times \frac{\kappa_{rj}}{1 + \mu^{-1}\kappa_{rj}} \qquad (7.7)$$

In this equation, $\Sigma_r'$ denotes a sum over only the reactive grains, and $\beta_{r0} = (\beta_r - \beta_{r1})$ is the equilibrium population of the unreactive states within the grain.

The eigenvalue is easily seen to have the following limiting behaviour: as $\mu_r \to \infty$, we recover the standard strong collision formula, equation (5.17); conversely, as $\mu_r \to 0$, the rate constant becomes strict Lindemann in form. Both limiting forms possess the same high pressure limit $\Sigma_i \tilde{n}_i d \equiv \Sigma_r \beta_r d_r$, but the two low pressure limits are vastly different: in the one case, the limit is $\mu \Sigma_r' \beta_r$, because any molecule excited into a reactive grain must eventually react but, in the other, the limit is $\mu \Sigma_i' \tilde{n}_i = \mu \Sigma_r \beta_{r1}$ because only those molecules excited directly into the reactive levels themselves can react.

The general behaviour of the rate constants for intermediate values of $\mu_r$ is exactly as shown in Figure 7.1 and 7.2 for the thermal isomerisation of methyl isocyanide; notice that for the cases where the randomisation processes are considered to be first order, a false high pressure limit appears, and that the true limit of $\Sigma_r \beta_r d_r$ is only achieved at much higher pressures. If the disparity between $\Sigma_r' \beta_r$ and $\Sigma_i' \tilde{n}_i$ becomes smaller, as it does for simpler molecules, then the false high pressure limit disappears, although its vestiges remain in the form of an inflection whose position depends upon the ratio of $\mu/\mu_1$; this effect is shown in Figure 7.8 for a model calculation [81.P2] imitating the thermal dissociation of carbon dioxide.

If the more general form, equation (7.2), for $\mu_r$, including both first and second order randomisation, is used and the analytic form, equation (7.7), is used to calculate the rate constants, the results found from the separable approximation in Figures 7.3–7.5 are recovered unaltered.

## 7.7 Directions for future development

We have just found that, for large molecules, the predicted behaviour is the same whether we use the separable or the non-separable model. Clearly, then, many situations where randomisation failure is thought to be a factor can be examined quite adequately by using the separable approach. One can envisage three such possible situations. Of obvious contemporary interest is the problem of multiphoton dissociation of large molecules by intense laser radiation; here, the model is readily extendable, conceptually, and we have shown how it can be used to find the reaction rate under steady illumination [82.P3; 82.V2] once the relevant randomisation rates and photochemical cross-sections have been allocated.

Also, as is becoming increasingly recognised [80.J2], multichannel thermal reactions possess considerable potential for showing up departure from strong collision behaviour; if such departures from strong collision behaviour arise as the result of a randomisation failure, this model quickly yields the relative fall-off behaviour for the various reaction channels, as an extension of equation (7.7). If there are only two interfering channels, $m = 1, 2$, the result is [81.P2]

$$\gamma_{0,m} \simeq {\sum_r}' \sum_{j=0}^{2} \frac{(D_m S_0, \bar{\rho}_{rj} S_0)}{1 + \mu^{-1} \kappa_{rj}} \qquad (7.8)$$

where $\kappa_{rj}$ are the three simple eigenvalues for grain $r$, and

Fig. 7.8. Fall-off curves for a model calculation on the thermal dissociation of carbon dioxide, considering only first order randomisation processes. In descending order, the curves correspond to $\mu_1 = \infty$, $10^{11}$, $10^{10}$, $10^9$, $10^8$, and $10^7 \, \mathrm{s}^{-1}$ and zero, respectively.

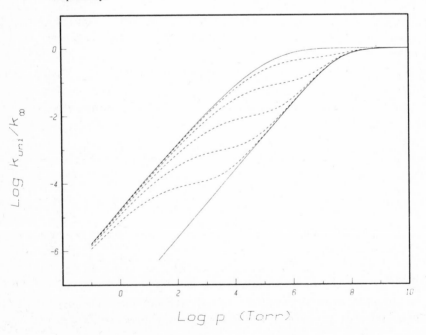

$$(D_m S_0, \bar{\rho}_{rj} S_0) = \frac{1}{Y} \times \frac{\beta_r}{\mu_r} \times \frac{\beta_{rm} d_m}{(\mu_r - \kappa_{rj} + d_m)}$$

$$Y = \frac{\beta_{r0}}{(\mu_r - \kappa_{rj})^2} + \frac{\beta_{r1}}{(\mu_r - \kappa_{rj} + d_1)^2} + \frac{\beta_{r2}}{(\mu_r - \kappa_{rj} + d_2)^2}$$

Third, and already well recognised, is the problem of decay versus stabilisation of a molecule initially prepared in a highly excited state, either photochemically or by a chemical activation process. The solution to this problem requires more than a knowledge of just the smallest eigenvalue [81.V2], and the obvious approach is to use the separable approximation.

### 7.8 Some further comments on the nature of randomisation processes

Examination of the nature of randomisation processes is still in its infancy, both on the experimental and theoretical fronts. The most straightforward experimental approach has been that of the formation of highly energetic symmetric molecules from unsymmetric precursors, followed by the observation of an asymmetric formation of products [71.R2; 81.L]. Next, we have many examples nowadays, e.g. [81.L; 82.S4], where non-statistical energy distributions are observed either in or between the products formed in unimolecular fragmentation reactions: these observations contain information about the degree of randomisation in the reactant molecule, but the deconvolution of such experimental data to obtain that information is likely to be a formidable task [81.P2]. Finally, we have a rapidly growing array of laser-based experiments which probe the dynamics of the molecular motions subsequent to the initial excitation, see e.g. [81.J].

On the theoretical front, it is possible to make a few simple assertions. We have already seen that a collisional component to the randomisation process may become faster the more dense are the states of the molecule. It is also obvious that the first order component will become slower as the states become further apart, but the molecular level density where this begins is not known; a cut-off at about 1000 states per wavenumber has been suggested [82.S2] for intramolecular vibrational relaxation of isolated molecules in one kind of experiment. It is also obvious that there must be propensity rules for the occurrence of randomising transitions within any grain [81.P2]: for example, transitions between states of similar angular momentum may be favoured.[5] There is already evidence from laser-excitation experiments on benzene [79.B2; 82.R1], and in Chapter 9 we will examine some quite old evidence for cyclopropane which may point in the same direction.

Also, equally obvious, is that in laser-induced processes, the extent of the energy range over which randomisation may take place will increase with increasing laser intensity because of the broadening effect of the laser field on the molecular states; to elaborate further would take us beyond the bounds of this book, but it is worth pointing out that many of the earlier so-called 'collisionless' experiments may not have been truly collisionless at all, either because of the presence of laser radiation or the presence of other molecules within the rather long range of the inter-molecular attractive potential.

5 Such a modification already falls within the framework of the present model – all that is necessary is to assume that the grains are defined by two criteria, the total internal energy, as before, as well as the manner in which that energy is allocated.

# 8

## Weak collision processes

In his discourse on the theory of unimolecular reactions, Bunker associated the onset of weak collision effects with the breakdown of either the strong collision or the randomisation assumptions [66.B2]. We have already constructed (in Chapter 7) an apparatus for examining the latter problem, and we will now attempt to inspect the problem of departures from the strong collision transition rates. By introducing the concept of a generalised strong collision, Nordholm and I have rendered the problem of detecting such departures a little more difficult. In the past, an internal relaxation rate lower than the hard sphere collision rate could be taken as, more-or-less, *prima facie* evidence for the weakness of the collisions, but as we have seen in the two preceding chapters, both cyclopropane (which conforms exactly to strong collision fall-off behaviour) and methyl isocyanide (which departs only very slightly from such behaviour) both appear to have internal relaxation rates of about one-tenth of the hard sphere collision rate. I have identified this so-called strong collision behaviour with an absence of dispersion in the rates of the normal-mode processes which are important in feeding the reactive states of the molecule. By this definition, collisions between helium and methyl isocyanide are probably *strong*, as we saw in Section 7.3, and so are collisions between argon and cyclopropane, as I will demonstrate in Section 9.4.

On the other hand, the strong collision treatment is quite poor in describing the shapes of the fall-off in rate with pressure for the reactions of many simpler molecules, as is shown for the case of the thermal dissociation of nitrous oxide in Figure 8.1: here, the experimental measurements [66.O] lie rather close to a strict Lindemann curve, whereas the strong collision shape exhibits a much more gradual decline. This approach to strict Lindemann behaviour is easily understood in terms of a sequential activation process: as the pressure declines and we enter the fall-off region, the states just above threshold decay so quickly that they cannot be replenished in full by activation processes from below; this is the normal fall-off mechanism. In the strong collision case, molecules are transferred directly into the higher reactive states from below threshold, whereas in a sequential activation process they must pass through those reactive states which lie in between; consequently, in the sequential process, relatively fewer molecules can be raised to higher reactive levels, and the fall-off with pressure becomes more pronounced. In the extreme, if the depopulation effect is severe enough, almost no molecules will progress beyond the lower part of the reactive region, and then we would approach the condition for strict Lindemann behaviour, that there is only one *effective* value of the decay rate constant. There is good numerical evidence from model calculations that this can happen [77.P]. Let us therefore examine the case of a tridiagonal relaxation matrix, which quite obviously can be made to represent a sequential activation process, and for which the unimolecular reaction rate problem is solvable analytically; in general, the eigenvalues of an arbitrary

Fig. 8.1. Comparison of the observed rates of dissociation of nitrous oxide at 2000 K with strict Lindemann and with strong collision behaviour. Notice that the limiting values of $k_{uni,0}$ and $k_{uni,\infty}$ used in constructing these curves are both about 8–9% higher than the values given by the original Arrhenius expressions in [66,O].

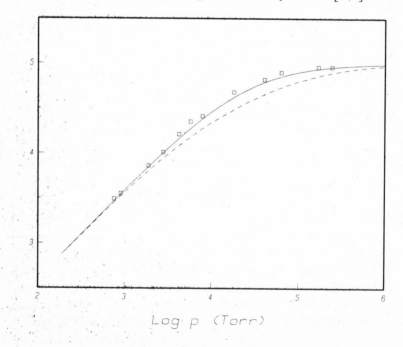

tridiagonal matrix will exhibit some dispersion and we may therefore, on the face of it, expect to find behaviour different from that caused by strong collisions which give rise to no such dispersion.

## 8.1 A tridiagonal reaction matrix

We divide up the energy-level spectrum of the molecule into grains, as usual, and we assign rate constants for transitions between adjacent grains only, i.e. $q_{i,i-1}$ and $q_{i,i+1}$; all other $q_{ij}$ ($j \neq i \pm 1$) are zero. We then form the relaxation matrix elements $[Q]_{ij}$ in the usual way by equation (2.4) and symmetrise it as in equation (2.10) or equation (3.3); let the elements of this symmetrised tridiagonal relaxation matrix be $[B]_{ij}$, where, in fact, only the entries $b_{i-1,i}$, $b_{ii}$, and $b_{i+1,i}$ are non-zero. The corresponding reaction matrix is simply $[B+D]$, and we are interested in $\gamma_0$, the smallest eigenvalue, which is the rate constant.

It is important to realise that the use of a tridiagonal reaction or relaxation matrix does not necessarily confine discussion to nearest-neighbour or step-ladder processes. Any non-degenerate symmetric matrix can be reduced to tridiagonal form by a similarity transformation, such as those associated with the names of Givens or Householder [65.W1]: thus, our tridiagonal reaction matrix can be regarded as being equivalent to a full reaction matrix, and if we only wish to examine the behaviour of the eigenvalues, then we will never know the difference; inconsistencies will arise, of course, if we wish to calculate some property which requires the eigenvectors, in which case we would have overlooked the Householder transformation in recovering the proper eigenvectors. This accounts for the fact that tridiagonal models have been so successful in the past, both in the numerical treatments of diatomic dissociation and of unimolecular reactions.

It is instructive to begin by examining the solution of $[B+D]$ when there is only one decaying state, which was first found by Yau [78.Y1; 78.Y3]; the presentation here follows the formulae given for the dissociation of the diatomic hydrogen molecule [79.Y1]. If there are $N$ grains labelled from $j=0$ to $j=N-1$, with only the latter being reactive, then the unimolecular rate constant can be written as

$$k_{\text{uni}} = [\text{M}] T^{-1} \tag{8.1}$$

where

$$T = \sum_{j=0}^{N-1} t_j \tag{8.2}$$

and

$$t_{N-1} = [\mathbf{M}]/d_{N-1}\tilde{n}_{N-1} \qquad (8.3)$$

otherwise

$$t_j = \left[\sum_{k=0}^{j} \tilde{n}_k\right]^2 / [(1-\tilde{n}_{N-1})\tilde{n}_j q_{j,j+1}] \qquad (8.4)$$

Although this is a neat expression, it is a little complicated to see through immediately. It has the property that it is strict Lindemann, as it must since there is only one reactive state; in addition, it possesses the bottleneck properties which one has to expect in a linear network [75.P1; 79.Y1]. If one of the $q_{j,j+1}$ is made very small, then the corresponding $t_j$ becomes very large, and the overall rate shrinks accordingly; if, on the other hand, all of the $t_j$ are of a similar magnitude, then the rate is determined collectively by all of them. The existence of a bottleneck in the activation process may be diagnosed fairly simply, as we have shown before [78.Y3; 79.Y1]: it occurs whenever we find a $t_j > t_{j+1}$, and the greater the disparity, the stronger is the bottleneck; the condition that $t_j > t_{N-1}$ corresponds, of course, to the fall-off region of pressure.

   If we now waive the restriction that there is only one decaying state in the system, and assign a different decay rate constant $d_r$ to every grain above threshold, a rather similar solution can be found. Our original result, stated in Theorem 4 of [81.V1], took the form of a recursive relation, with one pass required for each value of $d_r$; within each such recursion, bottleneck effects similar to those of equation (8.4) can be seen. More recently, Vatsya [82.V1] has developed a much simpler algorithm, giving upper and lower bounds to $\gamma_0$ (rather than to $\gamma_{ap}$ as given in [81.V1]), which takes the form

$$\frac{\lambda_1}{1+\lambda_1 M(x)} \leqslant \gamma_0 \leqslant \frac{1}{M(x)} \qquad (8.5)$$

$$M(x) = -\sum_{j=0}^{N-1} \frac{1}{J_{j+1}} \left[\sum_{i=0}^{j} \tilde{n}_i^{\frac{1}{2}} \prod_{l=i}^{j-1} \{B_{l,l+1}/J_{l+1}\}\right]^2.$$

$$J_1 = x - B_{00} - D_{00}$$

$$J_{i+1} = x - B_{ii} - D_{ii} - B_{i-1,i}^2/J_i$$

where $B_{ij}$ are the elements of the symmetrised relaxation matrix, $\lambda_1$ is its first non-zero eigenvalue, and $x$ is a starting approximation to $\gamma_0$. Under conditions where $\gamma_0$ is small compared with $\lambda_1$, $x$ can be taken to be zero, and the two bounds to $\gamma_0$ will still coincide to about four or five decimal places [82.V1]. No systematic examination of the bottleneck properties implicit in equation (8.5) has yet been attempted.[1]

## 8.2 Diagnosis of weak collision behaviour

The use of a tridiagonal relaxation or reaction matrix, in itself, does not guarantee weak collision behaviour: for although the symmetrised strong collision relaxation matrix, $\mu(1-p_0)$, cannot be reduced to tridiagonal form, matrices differing only minutely from it can. However, any thoughtfully constructed tridiagonal relaxation matrix may be expected to yield a fairly disperse set of eigenvalues, and so is likely to give a reasonable description of a weak collision process, as has been found on many occasions with step-ladder or nearest-neighbour models. The fundamental question we have to answer is whether the dispersion of the eigenvalues of the relaxation matrix is a necessary and sufficient condition for the occurrence of weak collision behaviour in unimolecular reactions. Or is there something more subtle?

Having dispensed with the notion that the observed internal relaxation rate $\mu$ discriminates between strong and weak collision regimes,[2] we are left with only one criterion for diagnosing a weak collision reaction, the occurrence of a fall-off curve which departs from that predicted by the strong collision expression, equation (5.14). We have seen two examples, that of methyl isocyanide in the previous chapter, and that of nitrous

---

1 No other analytic solution to the master equation for a weak collision system over the whole range of pressures has yet been found. A solution is known, at the low pressure limit only for a rather limited exponential probability model of a unimolecular reaction [77.T2; 80.F1], and Troe has developed empirical schemes for determining the pressure range over which the fall-off exhibits curvature and for joining smoothly the high and low pressure limiting solutions [77.Q; 79.T2].

   Notice also that Vatsya has provided a much more general solution for the rate of a unimolecular reaction in an intermediate regime, where the relaxation matrix is an arbitrary linear combination of a strong collision, equation (2.29), and a tridiagonal relaxation matrix [82.V1].

2 In fact, we cannot make an unambiguous definition of an apparent internal relaxation rate, $\mu$, because even if the low pressure limiting rate constant for the reaction is measured, its relationship to $\mu$ cannot be discerned until it is known what kind of bottleneck effect lies at the cause of the weak collision behaviour. For example, the two models described here give limiting values of $k_{uni,0}$ as $\mu\Sigma_r\beta_r$ (for a rate limiting activation bottleneck occurring at threshold) and $\mu\Sigma_r\beta_{r1}$ (for a severe bottleneck in the randomisation processes); since $\beta_r$ is the equilibrium population of the grain, whereas $\beta_{r1}$ is only the population of the reactive states within the grain, these two limiting rate constants are very different in general. Notice that in Troe's terminology [77.T2], $\beta_c$ would be unity for the first case, but $\beta_{r1}/\beta_r$ for the second.

oxide (and several other similar cases [78.Y4]) in Figure 8.1; the former was a fairly small effect, the latter a large one.

The next piece of evidence we have to consider is the almost universal insensitivity of calculated reaction rates when the transition probabilities in the model are varied; this can be seen in diatomic dissociation [75.P1], chemical activation [72.R; 77.Q], and in thermal unimolecular reactions [79.T2]. The reason for this is as follows. Since measurements are most often made at times long after the internal relaxation has ceased, the (normalised) steady distribution during the reaction is $(S_0)_i(\Psi_0)_i$, see equation (3.9). Moreover, the perturbed eigenvector $\Psi_0$ is rather similar to the unperturbed eigenvector $S_0$, with the dominant terms in the perturbation arising from the decay terms $d_r$. In fact, $\Psi_0 = (1 - \delta)S_0$, where

$$\delta \equiv \sum_{j=1}^{N-1} p_j \lambda_j^{-1} D^{\frac{1}{2}} \left[ 1 + D^{\frac{1}{2}} \sum_{j=1}^{N-1} p_j \lambda_j^{-1} D^{\frac{1}{2}} \right]^{-1} D^{\frac{1}{2}} \tag{8.6}$$

and $p_j$ is the eigenprojection $S_j(S_j,\ )$, assuming that $\gamma_0$ is small compared with all the $\lambda_j$ [81.P2]. Suppose that in a weak collision system we have two resolvable groups of processes, perhaps a rotational group and a vibrational group, whose mean rate constants might differ by (say) a factor of $10^2$: without knowing the actual magnitudes of the elements of the operator matrices $p_j$, other than that they are formed from normalised $S_j$, we can be sure that the group of processes with the larger $\lambda_j$ will make no effective contribution to $\delta$, and therefore to $\Psi_0$. Herein lies the germ of the explanation of the observed insensitivity of the computed rate constants to most of the elements of the relaxation matrix, and much refinement is possible, both analytically [72.B2] and by numerical experiment [71.M].

### 8.3 Bottlenecks as a cause of strict Lindemann behaviour

The only time when the calculated rate constants show any marked sensitivity to the variation of the elements of the relaxation rate matrix is when those elements happen to lie in the region of a bottleneck in the activation process [71.K1]. Thus, we might suppose, conjecturally, that marked deviations from strong collision fall-off behaviour will only occur when severe bottleneck effects are present in the activation processes.

We have already seen in the preceding chapter that if the randomisation processes present a bottleneck between the activation and the reaction steps, then the rate of reaction becomes more and more Lindemann in character as the severity of the bottleneck increases.

Can a similar effect be demonstrated for a bottleneck in the activation process? Vatsya [81.V1, Corollary 3] has shown that Yau's condition [78.Y4] for the occurrence of strict Lindemann behaviour as the result of a bottleneck in the activation ladder, although formally correct, is an unattainable one; his analysis of the tridiagonal problem, however, was unable to answer the question as to whether *almost* strict Lindemann behaviour could occur as the result of an activation bottleneck [81.V1]. The question can be settled (somewhat empirically, admittedly) in another way. Figure 8.2 shows the construction of a generalised weak collision relaxation matrix as a superposition of strong collision relaxation matrices with different rate constants. The area *abcd* represents a strong collision relaxation matrix connecting all states in the system with rate constant $\mu_0$: the area $a_1b_1cd_1$ depicts an additional strong collision relaxation matrix, connecting only states above a chosen level, with rate constant $\mu_1$; likewise, further additions $a_2b_2cd_2$, $a_3b_3cd_3$,... with rate constants $\mu_2$, $\mu_3$, ... up to as many terms as desired. One advantage of this matrix is that it has a very simple eigenvalue spectrum [81.V3], i.e.

Fig. 8.2. Schematic representation of a generalised weak collision relaxation matrix as superposition of strong collision relaxation matrices of different size and with differing relaxation rate constants.

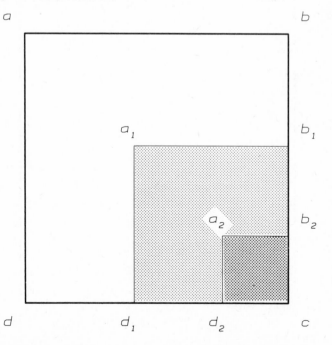

$\lambda_0 = 0$, $\lambda_j = \Sigma_{i=0}^{j} \mu_i$, whereas it is very difficult to control the eigenvalue spectrum of an arbitrary tridiagonal matrix without a lot of trial and error. It is obvious that if we choose $\mu_i > \mu_{i-1}$, then we can create a bottleneck in the activation process, and Figure 8.3 shows how increasing the severity of this bottleneck, positioned at threshold, in a model two-$\mu$ calculation on the methyl isocyanide reaction leads progressively to strict Lindemann behaviour; further details, with explicit formulae for the rate constant, are given in [81.V3]. Notice that the Lindemann form is a limiting one, and it is not possible to have a fall-off shape which curves more strongly than Lindemann [82.S3; 82.V3].

There is an urgent need for an analysis of bottleneck properties in collisional activation–deactivation processes, which is more general than equation (8.4), although (8.4) will probably be useful initially in characterising the position and the severity of the bottleneck. Whether or not there are other patterns of transition probabilities, not exhibiting bottle-

Fig. 8.3. Model calculation for the isomerisation of methyl isocyanide using a two-component generalised weak collision matrix as shown in Fig. 8.2. In ascending order, the curves are for $\mu_1 = 0$, $\mu_1 = \mu_0$, $\mu_1 = 10 \mu_0$, $\mu_1 = 100 \mu_0$, and $\mu_1 = \infty$; $\mu_1 = 0$ is the standard strong collision result, $\mu_1 = \infty$ is the strict Lindemann form.

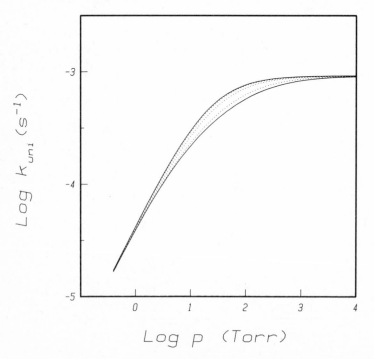

neck features (and ignoring the unphysical condition that the $k(E)$ is constant [81.V1]) which may lead to Lindemann behaviour, I am not sure, but I suspect not. Clearly, there is much work still to be done on this problem: some of this will have to take the form of systematic exploration of the properties of various transition probability matrices, but if a two-channel unimolecular reaction could be found for a molecule of intermediate size (4–6 atoms), a study of the relative fall-off shapes could be most instructive.

# 9

## *How well does it all work?*

The reader who has followed the development of this book from the beginning will recognise a progressive increase in the degree of speculation, chapter upon chapter; after describing the basic experimental phenomena, we began with a treatment of relaxation in simple molecules, which is virtually irrefutable, and ended up with an attempt to treat in a semiquantitative manner such concepts as those of randomisation and of activation bottlenecks, about which we still know really very little. For the sake of brevity, I have tried to keep the speculation to the minimum required to form a consistent foundation for the treatment of unimolecular reaction manifestations, wherever they may occur. My discussion has concentrated almost solely on the shape of the fall-off curve for a few thermal unimolecular reactions: this, despite the fact that I enumerated other interesting properties of simple thermal reactions in Chapter 1, and omitted entirely to mention the wide range of other experimental properties usually encompassed within the general topic of unimolecular reaction theory [72.R; 73.F]. In conclusion, therefore, I would now like to hold a brief inspection of each of those topics raised in Chapter 1, together with a few others so far not mentioned, to see how successful the present theory appears to be. By so doing, I hope that each such inspection will point the way to new and fruitful investigations: in some rapidly moving fields, such as those of multiphoton processes, or of pressure-dependent bimolecular reactions, comments made today may be totally out of date before these words are even read, but in relatively neglected areas, such as those of extremes of pressure, or of incubation and relaxation times, they may have a more lasting value.

### 9.1 The shape of the fall-off curve

Little needs to be added here concerning the shapes of fall-off curves. For the strong collision case, it was shown in Chapter 5 how such shapes are determined by an interplay between the way in which the Boltzmann population distribution and the specific rate function vary with energy in the reacting molecule. Moreover, the reproduction of the shape of the fall-off is marginally better if the purely empirical $k(E)$ function, derived as an inverse Laplace transform of the experimental rate law, is used rather than the semiempirical $k(E)$ function constructed according to the RRKM recipe. Likewise, it is rather better if the external rotational states are included in the required state counting than if they are excluded; however, the development of Chapter 6 provides powerful evidence for their inclusion.

Further, by introducing the postulate that the randomisation processes are no longer completely transparent, the well-documented departures from strong collision behaviour in the methyl isocyanide isomerisation are well reproduced, as shown in Chapter 7. Unfortunately, there is then a large gap in our experimental knowledge between this case and the traditional weak collision reactions, which exhibit almost pure Lindemann fall-off behaviour [78.Y4]. As a consequence, it is not possible to say for certain whether these weak collision effects arise from rate limitation through randomisation bottlenecks or through activation bottlenecks, or (more probably) through a combination of both; we have seen in the past two chapters that either form of bottleneck effect would yield its own limiting Lindemann form and, without the benefit of some unambiguous experimental data on intermediate cases, the choice must rely on intuition.

## 9.2 The position of the fall-off curve

Turning now to the position of the fall-off curve on the pressure axis, significant difficulties appear almost immediately. The fall-off is centred about the pressure for which the mean relaxation rate constant $\mu$ equals the mean Boltzmann averaged decay rate constant $\Sigma_r \beta_r d_r / \Sigma_r' \beta_r$ (Section 5.7). Because the shape of the fall-off curve (which we can calculate quite well) is determined directly by the dispersion of the function $\beta_r d_r$, it would seem that we can safely assume the Boltzmann averaged decay rate constant to be known and, therefore, that the observed centre of the fall-off defines the effective relaxation rate constant $r_i$. If, for the moment, we accept that the relaxation rate constant is *about* the same for all reactant molecules, the conclusion from Section 5.7 that the simpler the molecule, the larger is the numerical magnitude of the $k(E)$ function, immediately places any set of fall-off curves in the correct sequence, as found in Figure

1.1; thus, it is only when we come to a consideration of the precise numerical values of the $r_i$ that we get into trouble. Not the least problem now is that the theory has developed to the stage where small discrepancies in the temperature measurements of various works make detailed comparison between them difficult: we saw examples of this in discussing Figures 5.1, 5.8, 7.6 (perhaps also 8.1) and it will arise again below.

### 9.3 Apparent internal relaxation rates

#### (i) *Comparison with measured relaxation rates*

The apparent values of the internal relaxation rate constant $r_i$, deduced from the positions on the pressure axis of the unimolecular fall-off, were listed in Table 5.1 for six strong collision reactant molecules. There are, unfortunately, very little in the way of experimental data for the vibrational relaxation rates of these particular molecules, but application of the Lambert–Salter rule [77.L1] would suggest relaxation in 4–8 collisions for cyclobutane, for methyl isocyanide, and for ethyl isocyanide, all at room temperature, and perhaps in 20 collisions for ethyl chloride; such rates would not appear to be inconsistent with the deductions presented in Table 5.1. However, relaxation *measurements* have been made in the case of cyclopropane, and here there is a real disagreement: at room temperature, vibrational relaxation in cyclopropane takes about 1000 collisions, in good agreement with the Lambert–Salter prediction; in addition, Dove and Grant [80.D] have measured the relaxation rate from 340 to 430 K, and extrapolation of their data to the experimental temperature of 765 K suggests a relaxation rate constant of about $8 \times 10^3$ $Torr^{-1}s^{-1}$, a factor of 50–100 lower than the value required to account for the position of the fall-off in Table 5.1. A prerequisite for the understanding of the source of this discrepancy would seem to be the measurement of vibrational relaxation rates for as many as possible of the usual strong collision unimolecular reactant molecules *in or near* the kinetic temperature ranges.

#### (ii) *Variation of the apparent internal relaxation rate with temperature*

When we discussed the variation in the shape of the fall-off with temperature, in Chapter 5, we noted (but did not examine) the possibility of a small change in the rate constant $\mu = pr_i$, with temperature. Here, the results are quite disappointing for both the methyl isocyanide and the cyclopropane isomerisation reactions, the only two cases for which measurements exist. For methyl isocyanide, the apparent values of $\mu$ required to match the experimental data are $1.5 \times 10^6$, $1.2 \times 10^6$ and $1.5 \times 10^6$ Torr$^{-1}$ s$^{-1}$ at 473, 504 and 533 K respectively. It is unlikely that the true rate constants would vary in this irregular manner, but it would have required a temperature error of between 2 and 3 K to shift the $r_i$ value for 504 K from $1.2 \times 10^6$ to $1.5 \times 10^6$ Torr$^{-1}$ s$^{-1}$; given the care with which the temperatures were established in these experiments [62.S], this seems improbable. The situation is very similar for cyclopropane where there are three independent sets of measurements and therefore more scope for mutual inconsistencies. To match the observed data, one requires $r_i = 5.8 \times 10^5$ Torr$^{-1}$ s$^{-1}$ at 718 K [60.S], $5.5 \times 10^5$ Torr$^{-1}$ s$^{-1}$ at 765 K [53.P2] and $7.5 \times 10^5$ Torr$^{-1}$ s$^{-1}$ at 897 K [82.F2], a very similar irregularity. Unfortunately, the analysis in this case is complicated by the fact that the three corresponding values of $k_\infty$ lie on an Arrhenius plot which has a slope of about 71.5 kcal mol$^{-1}$, whereas the accepted value[1] is about 65.6 kcal mol$^{-1}$.

The fundamental problem here, of course, is that the rate constant $r_i$ is not a true observable quantity: it is derived by matching the observed and theoretical fall-off curves, and this matching is determined by the condition that

$$k_{\text{uni},0} = \mu \sum_r{}' \beta_r \qquad (5.16)$$

where $\Sigma_r' \beta_r$ is the total population, at equilibrium, of all states above threshold; in fact, to be more precise, it is the total equilibrium population of all states which are coupled to the reaction channels by the randomisation processes. By either definition, the summation, and therefore $\mu = pr_i$, is model dependent; hence, those uncertainties in vibration frequencies, anharmonicities, and the like which we dismissed in Section 4.3 may not be so unimportant after all. Also, the calculated value of $\mu$ is quite sensitive to the assumed value for the threshold energy of the reaction. Notice too that the arbitrary exclusion of any subset of states as being unimportant can have a marked effect on the calculated value of $r_i$: a simple example is to exclude the external rotational states in

---

1 If the three sets of data are brought into coincidence with an Arrhenius plot of slope 65.6 kcal mol$^{-1}$ by the arbitrary assumption that the real temperatures were 715, 765, and 912 K, the new values of $r_i$ become 7.2, 5.5, and $4.1 \times 10^5$ Torr$^{-1}$ s$^{-1}$ respectively, which some may find attractive, but which are hardly more convincing than the original set.

constructing $k(E)$, whence $r_i$ changes from $5 \times 10^5$ to $3 \times 10^6$ Torr$^{-1}$ s$^{-1}$ for the cyclopropane isomerisation at 765 K.[2]

It is a widely established practice to quote the average amount of energy transferred in a collision, again a model-dependent quantity: approximate equivalence with the present formalism is established by taking the product of $r_i/Z$ with the activation energy, on the grounds that this amount of energy is completely dissipated in $Z/r_i$ collisions. Recently, however, absolute measurements of the amount of energy transferred per collision have become possible [81.H], and it has been shown that toluene molecules excited to about 150 kcal mol$^{-1}$ lose about 0.4 kcal mol$^{-1}$ (150 cm$^{-1}$) per collision with helium and about 0.7 kcal mol$^{-1}$ (250 cm$^{-1}$) per collision with nitrogen; unfortunately, the position of the fall-off for the thermal decomposition of toluene in the presence of either of these gases is not known. Another similar experiment [82.R2] gives the internal relaxation rate constant for *tert*-butyl hydroperoxide as $8 \times 10^6$ Torr$^{-1}$s$^{-1}$; again, no fall-off measurements are available, but this relaxation rate is roughly an order of magnitude greater than those listed in Table 5.1.

### (iii) *Comparison with surface collision rates*

Information on the magnitudes of the internal relaxation rates is not limited, of course, to the observation of the position of the fall-off curve. A great deal of new information is being gathered these days by using various extensions of the original 'very low pressure pyrolysis' concept of Benson & Spokes [67.B]: the basic idea is to exploit the relationship, equation (5.21), but because the practical implementation takes the form of a flow experiment with mass-spectrometric sampling of the extent of reaction, the proper reduction of the experimental data requires a degree of sophistication [80.G2] beyond that which is appropriate here. The general result, as noted already in Chapter 1, is that gas–wall collisions are of the order of a few times more efficient than are gas–gas collisions in sustaining the unimolecular rate [73.G; 79.K1; 80.G2]; assuming that collisions with the wall have unit efficiency, then the general magnitude of the results listed in Table 5.1 appears to be amply confirmed.

The measurement of surface accommodation coefficients for the internal energy of gases has not been an easy task, but we can confidently expect to see significant advances in our understanding of such processes in the coming years, from a combination of techniques, for example

---

2 Without wanting to appear repetitive, it is clear from the build-up of the specific rate function shown in Section 6.2 that the external rotational states *must* be included.

vibrating surface measurements [81.R] or molecular beam scattering from surfaces [82.H]. The theoretical lines plotted in Figure 5.2 assume that the required accommodation coefficient for cyclopropane on glass is unity, and a recent very low pressure study suggests a value of between one-half and unity for the accommodation of chloroethane-2-$d_1$ on quartz near 1100 K [81.K2]. Likewise, the accommodation coefficient for *n*-octane on silica falls from about 0.66 at 350 to about 0.48 at 800 K [82.A]; a common theme emerging from almost all such experiments [e.g. 82.G; 82.Y2] is that the efficiency of the wall in transferring energy declines with increasing temperature.

### (iv) *An informative anomaly*

A revealing result was that found for the thermal isomerisations of *cis*-cyclopropane-$d_2$ and *trans*-cyclopropane-$d_2$ by Schlag & Rabinovitch [60.S]. The equilibrium *cis–trans* mixture is formed from either reactant at the same rate, and both substances undergo the standard structural isomerisation reaction to give propylene with equal rates; however, the geometric isomerisation is about 20 times faster than the structural isomerisation, although the activation energies for the two processes are indistinguishable.[3] The fall-off characteristics of both isomerisations are well represented by the simple strong collision formula, equation (5.14), and the shapes are indistinguishable; this must imply that over the range of energies sampled by the reaction at the temperature of the experiment, the two $k(E)$ functions for the respective processes remain in a constant ratio with each other – otherwise, the two fall-off curves would have different shapes. Also, just as expected, since the geometric isomerisation has a faster rate, its fall-off is at somewhat higher pressures: the values of $p_{\frac{1}{2}}$ are 1.5 and 5.6 Torr for the structural and geometric isomerisations respectively.

However, when we go on to calculate the internal relaxation rate constant, we find a different value for this quantity from each reaction.

---

3 The structural and geometric isomerisation reactions are not in competition with each other in the same sense as were the simultaneous reactions we touched upon in Section 5.9: if the *trans* molecule is taken as reactant, depletion of the population of a reactant grain by geometric isomerisation produces an equivalent population in a grain of the *cis* molecule, at the same energy, and with the same specific rate constant for decay to propylene; moreover, collisional relaxation of the newly formed product molecules in this grain to other states in the *cis* manifold is exactly compensated by collisional re-population of the depleted reactant grain (assuming that both isomers have similar relaxation behaviour). Thus, the structural isomerisation to propylene is unaffected by the presence of the geometric isomerisation processes and vice-versa.

The values of $r_i$ are about $6 \times 10^5$ Torr$^{-1}$ s$^{-1}$ (as before) for the isomerisation to propylene, and $4 \times 10^6$ Torr$^{-1}$ s$^{-1}$ for the interchange between
the *cis* and *trans* modifications. The difference between these two figures
is sufficiently large that it cannot be caused by an error in the experiment:
for both reactions to yield the same value of $r_i$, $p_{\frac{1}{2}}$ for the geometric
isomerisation would have to be near 30 instead of around 5 to 6 Torr!

It would appear that there are at least two distinct groups of states, one
set being the precursors for the ring-opening process and the other for
hydrogen–deuterium interchange. The primary relaxations into these two
manifolds of states are both 'strong' in the sense that the corresponding
groups of eigenvalues are unresolved, but their mean rates are different.
Randomisation within each manifold must be sufficiently fast for each
reaction to develop its full strong collision rate but, of course, these two
groups are not connected by the randomisation processes, otherwise the
distinction between them would be washed out. There is obviously scope
for a deeper examination of this result, especially in the light of two recent
observations: one is that excess energy deposited in one of four sub-
stituents of a tin atom is not shared throughout the whole molecule
before reaction [82.R3]; the other, that there may be two distinct
transition states in the bimolecular reaction of nitric oxide with ozone
[82.E].

## 9.4 Relaxation times and incubation times

In many shock-tube experiments on kinetic processes, such as the
measurement of the rates of ionisation of inert gas atoms [68.M], or of
dissociation of diatomic [65.W2; 69.W] or polyatomic [74.D] molecules,
we notice the appearance of what is usually called an induction time, or
an incubation time. In these experiments, the test gas is heated (virtually)
instantaneously from room temperature to the reaction temperature: the
reaction of interest does not commence immediately because it takes a
short period of time, roughly comparable with the internal relaxation
time, for the atoms or molecules to be excited to reactive states by
collision; the reaction begins gradually, and then accelerates quickly to
achieve a steady rate of reaction. A plot of the amount of product formed
v. the time from the arrival of the shock is shown schematically in Figure
9.1; if such a plot is extrapolated back linearly from the steady state rate
regime, the intercept on the time axis falls at a time later than the true
time zero, the difference being the incubation time $\tau_{inc}$. Once in the steady
regime, the reaction gives the appearance of having commenced at

$t_0 = \tau_{inc}$, and the amount of reactant remaining at time $t$ is

$$N(t) = \sum_i \eta_i(t) = N(0)e^{-k_{uni}(t - \tau_{inc})} \tag{9.1}$$

with $N(0) = \Sigma_i \eta_i(0) = 1$, by definition. Recalling equation (3.9)

$$N(t) = (\mathbf{S}_0, \mathbf{\Psi}_0)e^{-\gamma_0 t}(\mathbf{\Psi}_0, E^{-\frac{1}{2}}\eta(0)) \tag{3.9}$$

where the normalisations are that $(\mathbf{S}_0, \mathbf{S}_0) = (\mathbf{\Psi}_0, \mathbf{\Psi}_0) = 1$; remembering our new normalisation (introduced in Section 5.2), i.e. $(\mathbf{S}_0, \mathbf{\Psi}_0) = 1$, we now have to write

$$N(t) = (\mathbf{S}_0, \mathbf{\Psi}_0)e^{-\gamma_0 t}(\mathbf{\Psi}_0, E^{-\frac{1}{2}}\eta(0))/(\mathbf{\Psi}_0, \mathbf{\Psi}_0) \tag{3.9a}$$

whence separation of equation (9.1) into

$$N(t) = N(0)e^{-k_{uni}t} \times e^{k_{uni}\tau_{inc}} \tag{9.2}$$

gives immediately

$$k_{uni} = \gamma_0 \tag{9.3}$$

Fig. 9.1. Basis of the method for determining incubation times in chemical reactions.

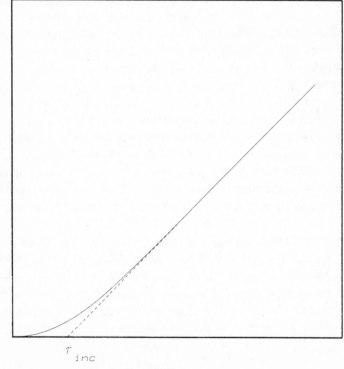

and

$$\tau_{inc} = k_{uni}^{-1} \ln[(\mathbf{\Psi}_0, E^{-\frac{1}{2}}\boldsymbol{\eta}(0))/(\mathbf{\Psi}_0, \mathbf{\Psi}_0)] \tag{9.4}$$

This derivation is a distillation from many given in [66.B1; 78.D; 79.K2; 79.Y2; 80.F1].

In the strong collision case, we may use equation (5.18) for $\mathbf{\Psi}_0$, whence

$$\tau_{inc} = \gamma_{ap}^{-1} \ln\frac{[\mu\Sigma_r\eta_r(0)/(\mu+d_r)]}{[(1-\gamma_{ap}/\mu)]} \tag{9.5}$$

Under conditions where the starting distribution has $\eta_r(0)=0$ for all reactive levels (as is the case in the traditional shock-wave heating experiment), the numerator inside the logarithm of equation (9.5) reduces to 1, whereupon (if $\gamma_0 \gg \mu$) $\tau_{inc}$ becomes $1/\mu$. Hence, in the strong collision case, the incubation time should be about equal to the internal relaxation time[4] whenever the reaction is initiated by a sudden heating of the cold gas; on the other hand, if $\boldsymbol{\eta}(0)$ is chosen to be the steady distribution $(S_0)_i(\mathbf{\Psi}_0)_i$, then $\tau_{inc}$ is zero, and if $\boldsymbol{\eta}(0)$ is chosen so that the reactive states are highly overpopulated, $\tau_{inc}$ can become negative.

Measurements of the incubation times exist for the isomerisation of cyclopropane, diluted to 1% in argon, at temperatures around 1200 K [73.D]. Unfortunately, I am not able to make an accurate reconstruction of the primary results from the data presented, but it is possible to discern the approximate magnitudes of both the relaxation time and the incubation time, for two temperatures, from two plots given in the paper.[5] At both temperatures (1179 and 1277 K), the graphical evidence indicates that the incubation time is indistinguishable from the relaxation time: certainly, the incubation period is not five times as long as the relaxation period, as it appears to be in some weak collision systems (see below). Thus, according to this evidence, the presence of a large excess of argon does not destroy the strong collision quality of the cyclopropane isomerisation; we saw a parallel result for helium and methyl isocyanide in Chapter 7.

In the weak collision case, although formally we can write an expression, equation (8.6), for $\mathbf{\Psi}_0$, there are no simple predictions, and we have to proceed by numerical experiment. For the only example for which extensive experimental data exist, the thermal dissociation of $N_2O$ [74.D], two rather different theoretical models lead, quite remarkably, to the same conclusion, unfortunately at some variance with the observed

---

4 Not zero, as has been suggested previously [80.F1].
5 The data plotted as squares give the incubation time and the data plotted as triangles give the relaxation time.

results. Yau and I [79.Y2] used a tridiagonal transition probability model to describe the activation processes in this weak collision system, and we obtained our results by direct computation; Forst & Penner [80.F1] used the exponential transition probability model, for which an analytic solution can be found at the low pressure limit. In each case, transition probability models were chosen in such a way that the inverse of the smallest relaxation eigenvalue was equal to the observed vibrational relaxation time $\tau_{rel}$ and four different transition probability models consistently gave the result that the ratio of $\tau_{inc}/\tau_{rel}$ should fall from around 2.5 at 2100 to around 2.0 at 3000 K; in the experiments this ratio falls from about 8 to about 5 over the same temperature range [74.D; 78.D].

A fallacy in these calculations, of course, is that of identifying the relaxation rate with the last eigenvalue $\lambda_1$ of the relaxation matrix: under conditions where the eigenvalues are too closely spaced to be resolved, the relaxation rate is an appropriate average of the rates of the contributing relaxations, see equations (2.20)–(2.23), and the observed rate will always be greater than $\lambda_1$; by how much we cannot know until we have as thorough an understanding of the polyatomic relaxation process as we now do for $H_2$. We can conclude, however, that there is no inconsistency between theory and experiment so far revealed by incubation measurements. For a much fuller discussion of the relationship between $\tau_{inc}$ and $\lambda_1^{-1}$ in weak collision systems, see [82.P4]; from this analysis, it would appear that the observation of $\tau_{inc} > \tau_{rel}$ in shock-heated reactions may become a useful indicator of weak collision behaviour as more experimental results become available.

## 9.5 The addition of inert gases

The extraction of information on relative collision efficiencies from measurements of the effect of adding non-reacting gas to the reaction depends upon the applicability of the so-called *mixture rule*: in the simplest case of a unimolecular reaction at its second order limit, this would mean that the total rate of reaction of A in the presence of inert B, C, ... would be

$$\text{Rate} = k_{uni,0} p_A + k_B p_B + k_C p_C + \dots \tag{9.6}$$

which is only strictly true if [77.B2] the reaction matrix $[A+D]$ commutes with $[B+D]$, $[C+D]$, ... ; this is an impossibly stringent requirement. Intuitively, departures from equation (9.6) are likely when the reaction is taking place under non-equilibrium conditions, as is the case here, but estimates of the expected extent of such departures are rather scant [70.M; 80.T], and they are probably small. There is one set of excellent experimental data on the isomerisation of methyl isocyanide in the presence of pentene-1, ethane, and helium [68.L], which shows strict adherence to equation (9.6), to within experimental error, in all three cases.

To understand what we should expect, consider the relaxation of A by A and by B, both according to the strong collision prescription, equation (2.26), with rate constants, $r_{i,A}$, $r_{i,B}$ respectively; since the two relaxation matrices commute, the mixture rule, equation (2.28), holds, and we can write

$$\mu = r_{i,A}p_A + r_{i,B}p_B = r_{i,A}(p_A + \alpha p_B) \qquad (9.7)$$

where $\alpha$ is the relative efficiency $r_{i,B}/r_{i,A}$; hence, if we were to plot the fall-off curve, equation (5.14), with the effective pressure $(p_A + \alpha p_B)$ on the horizontal axis, it would have the same shape as the fall-off curve for the undiluted reactant with $p_A$ as the ordinate.[6] There are a few cases in which a segment of a fall-off curve has been generated in this way, by using a fixed aliquot of reactant in the presence of varying pressures of the inert gas: methyl isocyanide[7] in the presence of propane, propylene, and hexafluoroethane [72.F4], trans-cyclopropane-$d_2$ [72.W], and 1,1-cyclopropane-$d_2$ [78.K], both in the presence of helium, but in no case can the results be interpreted to yield a change in fall-off shape, in the sense implied here.

It is only recently that departures from the mixture rule in relaxation have been found and, in fact, the effects are quite marked [77.G; 81.T]; the reasons for this behaviour are still quite poorly understood [79.P2; 81.M2], and appear to have escaped detection because they only show up at very high ratios of the inert gas to the relaxing gas concentrations. Of course, as I have argued in the discussion of Table 5.1, there is only, as yet, a tenuous connection between the observable relaxation rate constant for vibrational energy and the apparent relaxation rate constant $r_i$ required to define the position of the fall-off on the pressure axis, and so it is unclear at the present time whether a strong departure from one

6 Notice that for a strong collision system, the measurement of a change in rate with pressure is equivalent to a measurement of the change in $\mu$, no matter what part of the fall-off curve is being considered: it is not necessary to work at the low pressure limit to obtain an unambiguous value of the relative collision efficiency [70.C]; however, the change in rate is most sensitive to the change in pressure at the low pressure limit.

7 The case of methyl isocyanide and helium was discussed in some detail in Section 7.3.

mixture rule, equation (9.7), implies a like departure from the other, equation (9.6). When we are dealing with two weak colliders, then we can be sure that the appropriate pairs of matrices do not commute, so that both equations (9.6) and (9.7) will fail but, according to the numerical experiments of Troe [80.T], the effects 'are generally small and at the limit of sensitivity of today's experiments'.

The preceding paragraphs are rather finical, and they do not detract from the very large body of experimental collision efficiency data, mostly obtained by Rabinovitch and his students using the method illustrated in Figure 1.2, and which is still in need of proper interpretation at a much more fundamental level. It has seemed axiomatic that the way in which to understand the collision efficiency of an inert molecule in restoring the fallen rate of a unimolecular reaction was to try to calculate the rate of exchange of internal energy between the inert and the reactant molecule in a collision [70.L; 73.T; 77.N; 78.B2]; this exchange rate will always be faster than that for any exchange of internal with translational energy. However, our recent analysis of the relaxation of a mixture of two diatomic gases has led to the conclusion that the rate of relaxation of internal energy in the mixture is independent of the rate at which the two molecules exchange internal energy [81.M2]; if this conclusion carries across to the apparent internal relaxation rates which we require to understand the behaviour of the unimolecular reaction rate, then the calculation of these exchange rates is unnecessary. Assuming this is so, I have indicated elsewhere the direction in which the solution should aim [81.P1], but it seems that the general correlations of collision efficiency with measures of intermolecular forces will still be found [70.C; 77.T1; 82.G].

## 9.6 Weak collisions – which kind of bottleneck?

Strong collision behaviour is nothing more than a mathematical convenience which is never attainable in practice: there are two precise requirements for such behaviour, that the internal relaxation is pure exponential, equation (2.27), and that the rate of interchange between reactive and unreactive states above threshold is infinite. However, many thermal unimolecular reactions give the appearance of being strong collision processes, a fact which we can rationalise as follows. The internal relaxation is obviously not a pure exponential, but it mimics one moderately closely: for this to be so, there would not have to be any bottleneck in the relaxation process, which could happen if the rotational processes were unimportant, either because they were infinitely fast or because the rotational energy content is negligible in comparison with the vibrational energy content of the reacting gas. The unimportance of rotation in the relaxation process is not, of course, sufficient – we need also a pattern of vibrational transition probabilities which does not present an activation bottleneck. This is probably not a severe requirement in a polyatomic molecule, for we can already see the elements of this property in the relaxation of diatomic molecules: that because of anharmonicity, which crowds together the higher energy levels and increases their rate of interchange upon collision, the slowest vibrational relaxation process is the initial uptake of energy by the ground state molecule, see Figure 2.2 and the discussion in Section 2.4. Nor are the randomisation rates infinite but, as we saw in Chapter 7, they are often large enough to be considered so. Indeed, it becomes clear that strong collision behaviour is not, by and large, a co-operative property of the collision between the reactant and the other molecule, but it is an intrinsic property of the reactant molecule. It is the energy-level distribution of the reactant molecule itself which governs the degree of dispersion of the relaxation eigenvalues, and it is the same property which governs the rate of internal energy randomisation: it does not matter that the colliding partner is helium or argon, the shape of the fall-off curve is still to be found by using the strong collision formulae. The point at which the randomisation processes become rate limiting should, in fact, be relatively easy to locate through the measurement of specific rate functions for a series of *small* molecules and ions, using the techniques that already work well for larger systems, e.g. [79.R1; 82.B].

We can now see that both fundamental requirements for strong collision behaviour can fail [66.B2], and the problem is to decide which one has failed in any particular experimental situation; the basic apparatus now exists for us to be able to do this. We know that a bottleneck in the activation ladder will cause the fall-off curve to approach more nearly to the Lindemann shape. We know also that when the randomisation processes cease to be transparent, changes in shape occur in the fall-off curve; if one assumes a constant value for the randomisation rate constant for all grains, this also has the tendency to make the fall-off curves more Lindemann in shape. The next step, the effect of combining weak collision relaxtion properties with rate-limiting randomisation, has not yet been investigated: here, there is no easy formula such as (7.7) for the strong relaxation model, and progress will come initially by treating the relaxation and randomisation processes as being separable, as suggested in Chapter 7, also [80.P2]. As Just & Troe [80.J2] have pointed out, significant potential exists for learning about weak collision transition probability patterns from the study of multichannel decomposition reactions; the earliest numerical experiments of this kind were done on a sequential activation model by Chow & Wilson [62.C], but Just & Troe have given a formula which is valid at the low pressure limit for a special form of the collisional transition probability matrix. In fact, it is a trivial extension of equation (8.5) to compute the relative fall-off patterns for the whole pressure range of interest.

The other area of experiment which provides the opportunity to examine the interplay between weak collision effects and randomisation restrictions is that of chemical activation [72.R; 73.F], an area pioneered by Rabinovitch and his students [71.R2]. Despite musings to the contrary [81.V2; 81.P2],[8] it would seem that a proper treatment requires a theoretical framework which includes both weak collision relaxation processes and retardation due to the finiteness of the randomisation processes.

## 9.7 Other closely related topics

I have mentioned briefly, at earlier points, the topics of unimolecular reactions induced by intense radiation, of chemical activation, and of pressure-dependent bimolecular reactions and have ignored altogether those of the unimolecular decomposition of ions or of photochemically excited species; these are beyond the scope of a book as short as this one, although a few brief remarks about pressure-dependent bimolecular reactions may be helpful.

The simplest case is that of a recombination reaction, e.g.

$$CH_3 + CH_3 + M \rightarrow C_2H_6 + M$$

This is just the reverse of the dissociation reaction, for which our theories work quite well, and all that is necessary (at any pressure) to calculate the rate constant $k_r$ for the recombination from the rate constant $k_d$ for the dissociation is to make use of the rate quotient law $k_d/k_r = K_{eq}$, where $K_{eq}$

---

8 The usefulness of a strong collision relaxation matrix, but with a relaxation rate much less than the collision rate, is seen to wane upon examination of the true implication of equations (5.1) or (5.2): an excited molecule formed in a state a long way above threshold will not cascade down through other more weakly excited states before being stabilised – it is simply bled away, preponderantly to the lowest states in the system, at a steady rate; thus, this model fails to account for the significant curvature of reaction to stabilisation plots which occur at low pressures, often described as low pressure turn-up [63.K3].

is the equilibrium constant for the reaction; in general, the information required for the calculation of $K_{eq}$ is the same information as is required to attempt the fall-off calculations [78.T2]. It has been shown many times now that even though the reactions may be taking place (in the fall-off region) under non-equilibrium conditions, the rate quotient law is likely to hold to a very high degree of accuracy [61.R; 69.M2; 75.P1]. Thus, the collisional relaxation process which impedes the dissociation in one direction impedes the recombination to an equal extent when the reaction is performed in the opposite direction; the rate determining step, therefore, is not the association of the radical partners, but the collisional deactivation of the newly formed radical pairs.

It has become apparent in recent years that many simple bimolecular reactions are pressure dependent: an example of current topical interest is

$$ClO + NO(+M) \rightarrow Cl + NO_2(+M)$$

The simplest way to account for these reactions is to postulate the obvious intermediate, ClONO, to calculate its rate of formation just as one would calculate the rate of formation of ethane from methyl radicals, and to calculate its rate of dissociation to the products just as one would calculate the rate of dissociation of ethane to (say) ethyl radicals and hydrogen atoms. These reactions, therefore, are just a generalisation of the standard chemical activation process [74.L], but in which the 'stabilisation' product is not observed.

Another similar problem of very long standing is that of the disproportionation reactions of radicals: the prototype reaction may be taken to be

$$CH_3 + CH_3 \rightarrow CH_4 + CH_2$$

which, in fact, is not known although similar reactions in the fluorocarbon series are [81.C1; 82.N]. These reactions are really nothing more than the traditional chemical activation processes, and our understanding of them is seriously hampered by a lack of knowledge of the energy thresholds for the formation of the disproportionated products.

### 9.8 The transition state?

We have now arrived at the final paragraph of this essay without ever having had to invoke the concept of a transition state. The transition state has been central to our thinking about unimolecular reactions for over 50 years and has been an invaluable crutch upon which to support our intuition about the way in which chemical reactions take place. Unfortunately, its acceptance has led to a considerable overelaboration in the interpretation of experimental results; nevertheless, we cannot simply throw out the idea now without, at least, an inspection of the true meaning of the transition state.

Before doing this, it is worth observing the fact that some topics in chemical kinetics seem to depend less strongly on these concepts than do others; probably this is just a reflection of the inroads made by molecular dynamics into the description of the various phenomena. I have shown that it is no longer necessary even to think of a transition state in the traditional areas of thermal unimolecular reactions and also of chemical activation processes. The strongest impulse to invoke transition state ideas comes in our attempts to describe the very high rates of dissociation of molecules like ethane, and the reverse reaction, the recombination of methyl radicals: the universal notion of a 'loose' transition state possessing a lot of rotation [72.R; 73.F] parallels exactly my notion that the critical energy for reaction can be shared between the vibrational stretching motion and the overall rotation, as discussed briefly in Section 6.6, and the same is probably true for the unimolecular decomposition of ions. It is not surprising that current theoretical discussions of the disproportionation problem also devolve upon the attempt to formulate appropriate transition states different from those imagined to lead to the standard recombination reaction [79.M2]; nevertheless, these approaches, based as they are on the computation of the potential energy surfaces connecting the various chemical species, will be invaluable in helping to formulate the molecular motions which lead from the associated species to the products in question. This is exactly the same kind of information as we need to understand the relationship between the structural and the geometric isomerisations of 1,2 dideuteriocyclopropane, a problem we discussed earlier in this chapter.

We have not yet reached the point where we can say exactly what the transition state is, although in the corresponding problem of bimolecular exchange reactions, Kuppermann [79.K3] has found a set of conditions under which the Boltzmann average of the appropriate state-to-state cross-sections gives rise to the same rate constant as does transition state theory. Clearly, in a unimolecular reaction the transition state represents an ensemble average over a manifold of states at reaction energies, but it is not so clear which ensemble average we need. However, the evidence that we can glean from our consideration of such reactions as the geometric and structural isomerisation of cyclopropane, or from a comparison of the nature of dissociation versus disproportionation reactions, might indicate that the average is to be taken over those molecular states which are coupled to the particular reaction channel by the randomisation processes; we cannot account for the results of Schlag & Rabinovitch if we allow the average to be taken over a less specific ensemble.

Likewise, inspection of the approximate identity, equation (6.6), shows that the entropy of activation is a logarithm of an ensemble average for the microscopic decay rate constants, but a definitive examination of this correspondence would appear to be somewhat tricky.

Finally, I must come to the subject, raised in Chapter 1, of the behaviour of unimolecular reactions at very high pressures. At first sight, it is difficult to see how the transition state approach, which is apparently so successful in explaining the decline in rate with increasing pressure, could be supplanted by a molecular dynamic hypothesis. In fact, the transition state approach to the variation of chemical rates is beginning to show up inconsistencies, in the sense that there are now several examples of addition reactions in which $\Delta V\ddagger$ (of activation) is more negative than $\Delta V$ (of reaction) – in other words, the transition state is smaller than the final adduct [70.G2; 75.J; 80.L2]; this is nonsense! Thus, transition state theory as applied to high pressure reactions in general is beginning to show signs of failing, and it is probably just chance that it appears to work reasonably for the only two examples of unimolecular reactions so far studied with any certainty. The molecular dynamic approach to this problem will have as its principal ingredient the fact that the vibration frequencies of molecules are perturbed by pressure: the shifts that have been observed are often linear with pressure, they may be of either sign, and they are always small [81.I]; additionally, there can be cage effects at high densities, which cause a decrease in the anharmonicities of vibrations [81.M1]. This problem has yet to be explored, but when it is, I am confident that here, too, we will be able to simulate the experimental results without the need to resort to the construct of a transition state.

# APPENDIX 1

# *Units, symbols, and errata*

### A1.1 Units

Although many of the units used in this book conform to the standard SI practice, a few which are almost universal among practising kineticists do not: i.e. pressure in either atmospheres or in Torr, collision diameters in Å, and energy either in kcal mol$^{-1}$ or in wavenumbers (cm$^{-1}$). They are defined as follows:

Pressure  1 atmosphere (atm) $= 760$ Torr $= 101\,325$ Pa
Size       1 Å $= 10^{-10}$ m
Energy   1 kcal $= 4.184$ kJ $\equiv 349.76$ cm$^{-1}$

### A1.2 Symbols

Certain symbols are used throughout the text with a fixed meaning; others are defined locally and retain their meanings only within the context of that specific topic. The globally important symbols are as follows:

#### (1) *Standard chemical kinetic symbols*

$A$ and $E$    frequency factor and activation energy in Arrhenius expression
$A_\infty, E_\infty$    ditto, but at infinite pressure
$\beta$          inverse temperature, $1/kT$ or $1/RT$
$E, \varepsilon$       energy, in molar or in molecular units respectively
$E^*, \varepsilon^*$     threshold energy for reaction
$G(E)$      sum of states up to energy $E$
$g_i$          degeneracy of energy level $i$
$h$           Planck's constant
$k$           Boltzmann constant, or rate constant, as appropriate
$k_{\text{uni}}$       unimolecular rate constant in fall-off region, units s$^{-1}$

| | |
|---|---|
| $k_{uni,p}$ | ditto, at specified pressure $p$ |
| $k_0, k_\infty$ | abbreviations for $k_{uni,0}$, $k_{uni,\infty}$ respectively |
| $k(E)$ | specific rate constant for decomposition at energy $E$ |
| $\mu_n$ | reduced mass of nuclei |
| $\nu$ | molecular vibration frequency in wavenumbers |
| $p$ | pressure |
| $p_{\frac{1}{2}}$ | pressure at which $k_{uni}$ is half of $k_\infty$ |
| $Q(\beta)$ | partition function at inverse temperature $\beta$ |
| $Q(T)$ | ditto, at temperature $T$ |
| $\rho(E)$ | density of states at energy $E$ |
| $R$ | gas constant |
| $s$ | Kassel fall-off shape parameter |
| $t$ | time |
| $T$ | temperature |
| $\tau_{inc}$ | incubation time |
| $\tau_{rel}$ | relaxation time for internal energy |
| $Z$ | collision number |

## (2) *Operations and matrices*

| | |
|---|---|
| $\sum'$ | summation over reactive states or over grained states above reaction threshold |
| $(a,b)$ | scalar product of vectors $a$ and $b$ |
| $E$ | symmetrising operation, see equation (2.8) |
| $p_j$ | $j$th eigenprojection, defined by $S_j(S_j, )$ |
| $\mu(1-p_0)$ | symmetrised form of $A$ |
| $A$ | strong collision matrix, defined by equation (2.26) |
| $B$ | symmetrised form of weak collision (tridiagonal) matrix |
| $D$ | (diagonal) matrix of decay rate constants |
| $Q$ | general relaxation matrix, defined by equation (2.4) |

## (3) *Vectors*

| | |
|---|---|
| $S$ | (set of) eigenvectors of symmetrised form of relaxation matrix $Q$ |
| $S_0$ | eigenvector corresponding to zero eigenvalue, elements $(S_0)_i = \tilde{n}_i^{\frac{1}{2}}$ |
| $n(t)$ | vector of populations $n_i$, as function of $t$, in relaxation |
| $\Psi$ | (set of) eigenvectors of symmetrised form of reaction matrix $[Q-D]$ |
| $\eta(t)$ | vector of decaying populations $\eta_i$, as function of $t$, in reaction |

## (4) *Scalar quantities*

$\beta_r$     fraction of reactant molecules in grain $r$ at equilibrium

$\beta_{r1}$     ditto, in reactive quantum states

$\beta_{r0}$     ditto, in unreactive quantum states

$\gamma_i$     eigenvalues of matrix $[Q-D]$

$\gamma_0$     smallest eigenvalue of $[Q-D]$, $\equiv k_{\text{uni},r}$

$\gamma_{\text{ap}}$     zeroth iterative approximation to $\gamma_0$

$d_i$     decay rate constant for quantum state $i$

$d_r$     ditto, for average over grain $r$

$\lambda_i$     eigenvalues of relaxation matrix $Q$, $\lambda_0 = 0$

$\kappa_{rj}$     eigenvalues for randomisation and decay of grain $r$

$\mu$     internal relaxation rate, defined in equation (5.20), units $s^{-1}$

$\mu_r$     randomisation rate for grain $r$

$\mu_1, \mu_2$     first, second order randomisation rate respectively

$\tilde{n}_i$     fractional population in quantum state $i$, at equilibrium

$n_i$     ditto, in relaxation

$\eta_i$     ditto, in reaction

$N(t)$     total (decaying) population $(\Sigma \eta_i)$ at time $t$

        nb: $N(0) = \Sigma_i \tilde{n}_i = \Sigma_i n_i = \Sigma_r \beta_r = \Sigma_r (\beta_{r0} + \beta_{r1}) = 1$

$r_i$     second order rate constant for internal relaxation, units $\text{Torr}^{-1} s^{-1}$

$r_2$     rate constant for second order randomisation, units $\text{Torr}^{-1} s^{-1}$

### A1.3 Errata

The following misprints have come to my attention:

[78.Y2] Figure 10: the bottom label on the vertical axis is a factor of 10 too large.

[80.S] Equation (14): subscript $i$ should be $j$.

[81.V1] Page 412: first term (matrix $D$) missing from expression for $\phi(x) - \gamma$.

Page 417, Lemma 4: the second line should read

$$\eta_j = -\frac{1}{\mathscr{A}_{j,j-1}} \sum_{k=0}^{j-1} \beta_k, \quad \beta_k = (\chi_0)_k \{ \mu_k - (\chi_0)_k \sum_{l=0}^{n-1} (\chi_0)_l u_l \}.$$

Page 418, Theorem 5 and Corollary 5: in three places, the lower limit for summation should be changed from

$$j = v(m) \quad \text{to} \quad j = v(m) + 1$$

# APPENDIX 2

---

# *Rate constants for the thermal isomerisation of cyclopropane and for the thermal decomposition of cyclobutane*

Because of an editorial policy discouraging the presentation of experimental results in both graphical and tabular form, the primary rate constant data for the thermal isomerisation of cyclopropane in the fall-off region [53.P2] are only available in thesis form. One might have expected these results to have been superseded by now, but that has not happened, and Sowden's rather inaccessible thesis [54.S] remains the only source of these key data. In view of their continuing importance in the testing of unimolecular reaction theories, I am reproducing those results here (and also those of the cyclobutane reaction) for the convenience of future users.

Table A2.1. *R. G. Sowden, Ph.D. Thesis, Manchester, 1954*

Cyclopropane

| conversion (%) | time (s) | pressure (Torr) | temperature (°C) | $10^4 \times k_{uni}$ (s$^{-1}$) |
|---|---|---|---|---|
| 36.83 | 1799 | 34.0 | 489.2 | 2.55 |
| 58.71 | 3601 | 32.5 | 489.3 | 2.46 |
| 74.77 | 5401 | 31.8 | 489.9 | 2.55 |
| 37.23 | 2202 | 11.0 | 490.0 | 2.11 |
| 58.47 | 4860 | 5.00 | 489.9 | 1.81 |
| 43.11 | 7195 | 0.569 | 489.4 | 0.784 |
| 53.78 | 5402 | 0.982 | 495.3 | 1.43 |
| 57.03 | 10860 | 0.299 | 495.1 | 0.778 |
| 62.76 | 4505 | 3.59 | 495.9 | 2.19 |
| 35.43 | 2703 | 1.28 | 496.1 | 1.62 |
| 37.56 | 2380 | 2.45 | 495.8 | 1.98 |
| 70.04 | 4584 | 6.84 | 496.3 | 2.63 |
| 44.63 | 2152 | 6.38 | 497.3 | 2.75 |
| 39.00 | 2511 | 2.08 | 497.2 | 1.97 |
| 37.36 | 3604 | 1.37 | 491.0 | 1.30 |
| 40.23 | 2561 | 6.07 | 491.1 | 2.01 |

Table A2.1 (*continued*)

Cyclopropane

| conversion (%) | time (s) | pressure (Torr) | temperature (°C) | $10^4 \times k_{uni}$ (s$^{-1}$) |
|---|---|---|---|---|
| 66.67 | 7207 | 2.89 | 490.8 | 1.52 |
| 35.44 | 2522 | 4.04 | 490.5 | 1.73 |
| 36.81 | 4617 | 0.860 | 490.5 | 0.994 |
| 36.62 | 6440 | 0.370 | 490.7 | 0.708 |
| 34.44 | 6306 | 0.356 | 490.5 | 0.669 |
| 34.60 | 7150 | 0.245 | 490.4 | 0.594 |
| 68.07 | 8280 | 1.71 | 491.2 | 1.38 |
| 62.80 | 10804 | 0.634 | 491.0 | 0.915 |
| 35.08 | 7203 | 0.222 | 491.5 | 0.599 |
| 37.51 | 2823 | 3.29 | 491.4 | 1.66 |
| 36.04 | 5400 | 0.495 | 492.0 | 0.827 |
| 36.88 | 5880 | 0.430 | 491.7 | 0.782 |
| 38.17 | 9180 | 0.188 | 491.4 | 0.523 |
| 32.56 | 9840 | 0.120 | 491.4 | 0.401 |
| 32.57 | 10920 | 0.094 | 491.4 | 0.361 |
| 33.66 | 8160 | 0.170 | 491.6 | 0.503 |
| 35.07 | 10140 | 0.103 | 492.2 | 0.426 |
| 38.74 | 15360 | 0.067 | 491.9 | 0.319 |
| 38.61 | 10200 | 0.152 | 491.7 | 0.478 |
| 35.38 | 9900 | 0.134 | 492.0 | 0.441 |
| 39.59 | 2106 | 14.7 | 492.0 | 2.39 |
| 53.22 | 2414 | 84.1 | 492.0 | 3.15 |

Cyclobutane

| conversion (%) | time (s) | pressure (Torr) | temperature (°C) | $10^4 \times k_{uni}$ (s$^{-1}$) |
|---|---|---|---|---|
| 35.16 | 1262 | 5.02 | 448.3 | 3.43 |
| 32.14 | 2000 | 0.263 | 448.3 | 1.94 |
| 21.95 | 1801 | 0.102 | 448.3 | 1.37 |
| 33.83 | 1197 | 7.70 | 448.8 | 3.45 |
| 34.32 | 1492 | 1.47 | 448.4 | 2.82 |
| 37.73 | 1792 | 1.01 | 448.3 | 2.64 |
| 32.24 | 1580 | 0.720 | 448.2 | 2.46 |
| 26.99 | 1759 | 0.237 | 448.2 | 1.79 |
| 42.88 | 1698 | 3.49 | 448.3 | 3.30 |
| 16.19 | 1445 | 0.067 | 448.3 | 1.22 |
| 17.73 | 1850 | 0.057 | 448.3 | 1.06 |
| 28.31 | 1640 | 0.323 | 448.3 | 2.03 |
| 29.37 | 1592 | 0.420 | 448.3 | 2.18 |
| 42.85 | 1730 | 2.27 | 448.4 | 3.23 |
| 22.55 | 1460 | 0.185 | 448.6 | 1.75 |
| 17.49 | 1603 | 0.063 | 448.6 | 1.20 |
| 53.46 | 3594 | 0.388 | 448.6 | 2.13 |
| 21.25 | 1797 | 0.088 | 448.6 | 1.33 |

Table A2.1 (*continued*)

Cyclobutane

| conversion (%) | time (s) | pressure (Torr) | temperature ( C) | $10^4 \times k_{uni}$ (s$^{-1}$) |
|---|---|---|---|---|
| 22.26 | 1557 | 0.140 | 448.8 | 1.62 |
| 20.41 | 1521 | 0.125 | 448.6 | 1.50 |
| 48.97 | 3204 | 0.371 | 448.6 | 2.10 |
| 55.38 | 2962 | 1.06 | 448.4 | 2.72 |
| 39.16 | 3530 | 0.101 | 448.4 | 1.41 |
| 50.47 | 3136 | 0.451 | 448.6 | 2.24 |
| 60.00 | 3250 | 1.61 | 448.3 | 2.82 |
| 31.85 | 1586 | 0.587 | 448.3 | 2.42 |
| 19.61 | 1750 | 0.074 | 448.6 | 1.25 |
| 37.83 | 1459 | 3.25 | 448.6 | 3.25 |
| ∼30 | — | 29.5 | 448.4 | 3.88 |
| ∼60 | — | 61.7 | 447.5 | 3.42 |
| ∼30 | — | 93.3 | 447.3 | 3.31 |
| ∼60 | — | 102.0 | 447.0 | 3.31 |

In the last four experiments, the rate constant was deduced from a plot of the total pressure versus the time.

Table A2.2 *A. D. Kennedy, Ph.D. Thesis, Manchester, 1963*

Cyclopropane
in a 1 litre sphere

| ratio[a] | time (s) | pressure (Torr) | temperature (K) | $10^4 \times k_{uni}$ (s$^{-1}$) |
|---|---|---|---|---|
| 7.47 | 8250 | 6.7 | 770.8 | 2.96 |
| 0.701 | 2700 | 6.6 | 765.0 | 2.25 |
| 1.30 | 3823 | 5.5 | 771.7 | 2.49 |
| 0.763 | 2685 | 4.6 | 772.2 | 2.41 |
| 0.731 | 5495 | 1.5 | 763.1 | 1.14 |
| 0.571 | 10270 | 0.23 | 763.5 | 0.501 |
| 0.147 | 14940 | 0.0089 | 762.5 | 0.105 |
| 0.356 | 22500 | 0.016 | 764.4 | 0.162 |
| 0.770 | 85500 | 0.0018 | 764.4 | 0.080 |
| 0.411 | 26220 | 0.028 | 764.4 | 0.157 |
| 4.77 | 89100 | 0.040 | 764.4 | 0.235 |
| 0.580 | 56700 | 0.0040 | 764.5 | 0.096 |
| 2.00 | 154800 | 0.00066 | 764.5 | 0.085 |
| 1.73 | 63000 | 0.026 | 764.6 | 0.189 |
| 1.96 | 172800 | 0.0010 | 764.6 | 0.076 |

## Table A2.2 (*continued*)

Cyclopropane
in a 1 litre sphere

| ratio[a] | time (s) | pressure (Torr) | temperature (K) | $10^4 \times k_{uni}$ (s$^{-1}$) |
|---|---|---|---|---|
| packed vessel | | | | |
| 1.12 | 22500 | 0.059 | 771.0 | 0.414 |
| 0.808 | 13980 | 0.092 | 771.0 | 0.526 |
| 4.74 | 56100 | 0.023 | 771.1 | 0.386 |
| 6.36 | 74700 | 0.011 | 771.1 | 0.331 |
| 3.52 | 62500 | 0.0030 | 767.0 | 0.286 |
| 2.00 | 49680 | 0.00065 | 769.8 | 0.278 |
| 2.62 | 66600 | 0.0053 | 764.0 | 0.236 |

[a] ratio = observed ratio of propylene–cyclopropane, without dead-space correction

# APPENDIX 3

# *Computer programs for thermal unimolecular reactions*

This appendix provides a set of self-contained computer programs for the calculation of rate constants in thermal unimolecular systems. Most of them were written by Andrew Yau [78.Y2], and have now stood up to five years, or so, of continual use. They may run into trouble occasionally if anharmonicity values in the range 0.005–0.006 are used for several oscillators at the same time, or if other unusual molecular parameters are tried but, by and large, they perform acceptably in normal circumstances. Some effort has been expended in trying to make the repetitive segments of the calculations efficient, but there is no pretence at elegance – a contradiction in terms when one is constrained to use FORTRAN [72.D3].

The organisation of the routines is as follows. The MAIN routine handles the flow of the calculation and the assembly of the input parameters. Then there is a UTILITY package which calculates the required sums and densities of states with appropriate precision; this is sufficiently efficient that it is no longer necessary to group the molecular frequencies in most cases. It also divides the reactive energy range into grains of the specified size, computes the equilibrium grain populations $\beta_r$, and their respective specific rate constants $d_r = k(E)$ by the inverse Laplace transform method, equation (4.9). Finally, UNIRAT calculates the rate constant at the requested values of the pressure. As the nature and quantity of output is a matter of taste, all output instructions have been removed (but convenient locations for printing are noted by remarks); by this device, I can ameliorate some of the ugliness by omitting the corresponding FORMAT statements.

With the notations provided, the MAIN and UNIRAT routines should be easy to follow, but the UTILITY package requires some better description. Basically, it provides a means of calculating the required density of states by direct count up to a predefined energy, followed by a steepest descent approximation up to the maximum energy specified in the input data. Descriptions of each routine follow.

COUNT: The vibrational state degeneracy $g_v(\varepsilon)$ is computed from $\varepsilon = 0$ to $\varepsilon_m$ by direct count.

First, the oscillators are arranged in descending order of their fundamental frequencies. Then, for each oscillator, labelled $j$, all the energy levels $\varepsilon_{ij}$ up to $\varepsilon_m$ are calculated, with $\varepsilon_{0j} = 0$ and $\varepsilon_{\max,j} \leqslant \varepsilon_m$. An array of energy states for the first pair of vibrators, up to $\varepsilon_m$, is then constructed by convoluting the energies of the second oscillator with those of the first oscillator, viz. ($\varepsilon_{02} + \varepsilon_{01}$, $\varepsilon_{02} + \varepsilon_{11}$, $\varepsilon_{02} + \varepsilon_{21}$, ... $\varepsilon_{02} + \varepsilon_{\max,1}$; $\varepsilon_{12} + \varepsilon_{01}$, $\varepsilon_{12}$ $+ \varepsilon_{11}$,...). Each of the energy levels for the third oscillator is then convoluted with this combined array of states, forming a new combined array, and so on until all of the oscillators have been included. These arrays terminate at $\varepsilon_m$, and grids of width $\delta\varepsilon$ are formed: the state degeneracy of the $i$th grid is taken to be the sum of the $g_{i,v}$ between $(i-1)\delta\varepsilon$ and $i\delta\varepsilon$.

This procedure is similar to the Stein–Rabinovitch method [73.S], but differs from it in some important respects. Counting is performed only at energies where an energy state is present, which results in an appreciable saving in time when $\varepsilon_m$ is not high (as is the case here where $\varepsilon_m$ is usually a few thousand wavenumbers); this is because there are only a few states near $\varepsilon = 0$, and one is therefore performing counting and convolution only over these states, and not at all the energy grids as in the Stein–Rabinovitch method. Also, the convolution is performed before the energy levels are rounded, thereby reducing to a minimum (with little additional programming effort) the propagation of round-off errors in the subsequent convolution with the rotational state sum.

EXTROT: The rotational state sum $G_r(\varepsilon)$ is calculated up to $\varepsilon_m$ and convoluted to obtain the vibration–rotation state sum $G_{vr}(\varepsilon)$.

The rotational state degeneracy $g_r(\varepsilon)$ for linear and symmetric top molecules may be evaluated from standard formulae [53.P1; 62.D]. The rotational state sum is then

$$G_r(\varepsilon) = \sum_{\varepsilon} g_r(x) \qquad\qquad (A3.1)$$

where the summation refers to all energies up to and including $\varepsilon$. The quantity $G_r(\varepsilon)$ is calculated at intervals of $\Delta\varepsilon$ (where $\Delta\varepsilon$ is the grain size for the final rate constant calculation, an integral multiple of the grid size $\delta\varepsilon$), and it is then convoluted with $g_v(\varepsilon)$ to give

$$G_{vr}(\varepsilon) = g_v {}^* G_r \qquad\qquad (A3.2)$$

The convolution procedure starts from $\varepsilon_m$ and proceeds *down* to $\varepsilon = 0$ (as opposed to the conventional way of starting upwards from $\varepsilon = 0$) because this results in a saving of half the required storage space. The state sum up to the $i$th grid may be written as

$$G_{i,vr} = G_{vr}(\varepsilon = (i-1)\Delta\varepsilon) \tag{A3.3}$$

RINCNT: If there are any internal rotational degrees of freedom, the state degeneracy $g_i(\varepsilon)$ of the internal rotation is convoluted with $G_{vr}(\varepsilon)$ to obtain the sum of internal states $G(\varepsilon)$ of the molecule. Again, this convolution is performed downwards.

DESENT, option NFLAG $= 1$: The sum of states from $\varepsilon = \varepsilon_m + \Delta\varepsilon$ to $\varepsilon = \varepsilon_{max}$ is calculated at intervals of $\Delta\varepsilon$ by a non-iterative steepest-descent procedure [77.Y2].

The sum of states at energy $\varepsilon$ is given by

$$\left.\begin{array}{l} G(\varepsilon) = \{Q(\beta)e^{\beta\varepsilon}\beta^{-1}[2\pi\xi(\beta)]^{-\frac{1}{2}}\}|_{\beta^*} \\[2mm] \xi(\beta) = \partial\zeta/\partial\beta \quad \text{and} \quad \zeta(\beta) = \dfrac{\partial\ln Q}{\partial\beta} - \beta^{-1} \end{array}\right\} \tag{A3.4}$$

Equation (A3.4) is evaluated at $\beta^*$, the saddle point, equal to the inverse temperature at which the internal energy

$$\varepsilon(\beta) = -\frac{\partial\ln Q}{\partial\beta} = \varepsilon - \beta^{-1} \tag{A3.5}$$

The appropriate value of $\beta^*$ is easily related to an initial guess $\beta_0$ by

$$\beta^* = \beta_0 - \Delta\varepsilon\left\{\xi^{-1}\left(1 - \tfrac{1}{2}\Delta\varepsilon\xi^{-2}\frac{\partial\xi}{\partial\beta}\right)^{-1}\right\}\bigg|_{\beta_0} \tag{A3.6}$$

where

$$\Delta\varepsilon = \varepsilon + \zeta(\beta_0) \tag{A3.7}$$

and the quantity $\zeta(\beta_0)$ is defined as in equation (A3.4). Note that $\beta_0$ and $\zeta(\beta_0)$ are just the $\beta^*$ and (minus) the $\varepsilon$ value of the preceding energy, except for the first cycle, where $\beta_0$ is estimated by using the following empirical relation

$$\beta_0 = \frac{r+2}{2\varepsilon} + \frac{m}{S_m}\ln(1 + S_m/\varepsilon) \tag{A3.8}$$

where $r$ is the number of one-dimensional plus twice the number of two-dimensional rotations and

$$S_m = \sum_{i=1}^{m} g_i v_i \tag{A3.9}$$

with $m = 1$ if $g_1 v_1 \geqslant \varepsilon$, $m = n$ if $S_n \leqslant \varepsilon$, and otherwise $S_m \leqslant \varepsilon < S_{m+1}$ [77.Y2].

This routine requires the existence of FNPHI1 or FNPHI2, which calculate the average internal energy and its first two derivatives with respect to $\beta$; the former uses exact partition functions for the Morse oscillators, whereas the latter uses approximate ones, with a corresponding saving of almost an order of magnitude in the computing time.

DESENT, option NFLAG = 0: The density of states $\rho(\varepsilon)$ is calculated from $\varepsilon_\infty$ to $\varepsilon_\infty + \varepsilon_{max}$ by the same procedure.

We can write

$$\rho(\varepsilon) = \left\{ Q(\beta) e^{\beta \varepsilon} \left[ 2\pi \frac{\partial^2}{\partial \beta^2} \ln Q \right]^{-\frac{1}{2}} \right\}\Bigg|_{\beta\dagger} \tag{A3.10}$$

where $\beta\dagger$ is the inverse temperature at which the internal energy $\varepsilon(\beta) \equiv \partial \ln Q / \partial \beta = \varepsilon$, and is given by

$$\beta\dagger = \beta_0 - \Delta\varepsilon \left\{ \left( \frac{\partial^2}{\partial \beta^2} \ln Q \right)^{-1} \left[ 1 - \tfrac{1}{2}\Delta\varepsilon \left( \frac{\partial^2}{\partial \beta^2} \ln Q \right)^{-2} \left( \frac{\partial^3}{\partial \beta^3} \ln Q \right) \right]^{-1} \right\}\Bigg|_{\beta_0} \tag{A3.11}$$

where $\Delta\varepsilon = \varepsilon + \partial \ln Q / \partial \beta|_{\beta_0}$. Again, $\beta_0$ and $\partial \ln Q / \partial \beta|_{\beta_0}$ are just the $\beta\dagger$ and (minus) the $\varepsilon$ value corresponding to the preceding energy except for the first cycle, where a sufficient approximation is

$$\beta_0 = \frac{r}{2\varepsilon} + \frac{m}{S_m} \ln(1 + S_m/\varepsilon) \tag{A3.12}$$

A much fuller description of these procedures is given in the original publication [77.Y2].

At this juncture, the MAIN routine is given the temperature at which it is required to calculate the rate, whereupon it proceeds to calculate the molecular partition function for this temperature with PFNCTN. Hence, we now have

$$G_i \equiv G(\varepsilon = (i-1)\Delta\varepsilon), \qquad 1 \leqslant i \leqslant i_{max}$$
$$\rho_i \equiv \rho(\varepsilon = \varepsilon_\infty + (i-1)\Delta\varepsilon), \quad 1 \leqslant i \leqslant i_{max}$$

whence it is possible to proceed with the final stages in the calculation.

GRAIN: The reactive energy range ($\varepsilon_\infty$ to $\varepsilon_\infty + \varepsilon_{max}$) is grained into channels, with states between $\varepsilon_r$ and $\varepsilon_{r+1}$ being grouped into the $r$th channel, the equilibrium populations $\beta_r$, and the specific decay rate constants $d_r$, are calculated. It is not necessary for these grains to be of constant width (in energy) but, for convenience, they are in this program; the number of grains is specified by the parameter NGRAIN.

The population of the $r$th pseudo-level is given by

$$\beta_r = \{Q(\beta)\}^{-1} \int_{\varepsilon_r}^{\varepsilon_{r+1}} \rho(\varepsilon) e^{-\beta \varepsilon} d\varepsilon$$
$$\simeq \{Q(\beta)\}^{-1} \Delta\varepsilon \left\{ \sum_i \rho(\varepsilon_i) e^{-\beta \varepsilon_i} - \tfrac{1}{2}[\rho(\varepsilon_r) e^{-\beta \varepsilon_r} + \rho(\varepsilon_{r+1}) e^{-\beta \varepsilon_{r+1}}] \right\} \tag{A3.13}$$

where the summation over $i$ refers to the grid points from $\varepsilon_r$ to $\varepsilon_{r+1}$.

For the calculation of $d_r$ in the 'discrete region', i.e. when $\Delta\varepsilon \times \rho(\varepsilon) \leqslant 1$, equation [55] of [78.Y2] is used, viz.

$$d_r = A_\infty \{Q(\beta)\beta_r\}^{-1} e^{-\beta\varepsilon_\infty} \sum_i g_i e^{-\beta\varepsilon_i} \tag{A3.14}$$

In the 'continuous region', i.e. when $\Delta\varepsilon \times \rho(\varepsilon) \gg 1$, the integral form is used, viz.

$$d_r = A_\infty \{Q(\beta)\beta_r\}^{-1} e^{-\beta\varepsilon_\infty} \int_{\varepsilon_r - \varepsilon_\infty}^{\varepsilon_{r+1} - \varepsilon_\infty} \rho(\varepsilon) e^{-\beta\varepsilon} \, d\varepsilon$$
$$\simeq A_\infty \{2Q(\beta)\beta_r\}^{-1} (1 + e^{\beta\Delta\varepsilon}) e^{-\beta\varepsilon_\infty} \sum_i (G_{i+1} - G_i) e^{-\beta\varepsilon_i} \tag{A3.15}$$

where the $G_i$ are as defined above, and the summation is over grid points from $\varepsilon = \varepsilon_r - \varepsilon_\infty$, to $\varepsilon = \varepsilon_{r+1} - \varepsilon_\infty$; the factor $(1 + e^{\beta\Delta\varepsilon})/2$ gives an acceptable approximation to the Boltzmann factor over the range $\Delta\varepsilon$.

```
C     THE MAIN ROUTINE (1) directs the flow of the program, and
C                       (2) handles the input of data for the
C                           calculation of state-sum, state-degeneracy
C                           and reaction rate of the molecule.

      REAL*4 E(25,100),PCONC(100),UNIK(100),ELEVEL(4000),P(4000)
      REAL*4 WE(25),XE(25),WEXE(25),G(25),GWE(25),WEXE2(25)
      REAL*4 BINROT(5),EPRINT(5),PPRINT(5),TITLE(16)
      INTEGER IG(25),NMAX(25),IE(4000)
      COMMON/BLK1/P
      COMMON/BLK2/ELEVEL,IE
      COMMON/BLK3/G,GWE
      COMMON/BLK4/E,NMAX
      COMMON/BLK5/WE,IG
      COMMON/BLK6/XE
      COMMON/BLK7/WEXE2
      COMMON/BLK8/WEXE
      COMMON/BLK9/PCONC,UNIK
      COMMON/BLK10/BINROT
      COMMON/BLK12/RINF,RANDK1,RANDK2,WCRIT
C     NDIM is size of arrays P, IE and ELEVEL
      NDIM=4000
      FDIM=FLOAT(NDIM)
      NDIM2=NDIM/2
      READ(5,#) (TITLE(I),I=1,16); fewer than 72 characters.
      READ(5,#) AFACTR,ECRIT
CARD#2    AFACTR: preexponential factor (sec-1) in ARRHENIUS equation;
C         ECRIT: activation energy at infinite pressure (kcal/mol).
      ECRIT=ECRIT*3.497591E2
      READ(5,#) NPRINT,NCONC,NCHANL
CARD#3    NPRINT,NCONC,NCHANL
C         OPTIONS:  program prints the state-sum and state-density
C         of the molecule at every NPRINT energy;
C         no printout if NPRINT=0.
C         NCONC is number of pressures to be used (NCONC < 100);
C         NCHANL is number of reactive pseudo-levels; =1 for
C         LINDEMANN fall-off; = 10-15 for proper convergence.
      READ(5,#) NOSC,NINROT,NEXROT,NMORPF
CARD#4    NOSC is number of DIFFERENT oscillator frequencies
C         in the molecule (<25);
C         NINROT is number of 1-dimensional internal rotations (<5);
C         NEXROT is number of external rotation; =1 for Z-rotation,
C         =2 for XY-rotation, =3 for symmetric top or spherical top;
C         NMORPF - type of oscillator: =0 for harmonic oscillators,
C                  =1 for Morse oscillators, =2 for Morse osc. with
C                  approximate partition function.
      IF (NMORPF-1) 5,2,4
    2 CONTINUE
C*** Print flag showing that exact partition functions were used.
      GO TO 5
    4 CONTINUE
C*** ditto, approximate partition functions.
    5 NROT=NINROT+NEXROT
      ROTINT=1.0E0
      IF (NINROT) 21,21,6
    6 READ(5,#) (BINROT(I),I=1,NINROT)
CARD#5    (If NINROT > 0)  (BINROT(I),I=1,NINROT): rotational
C         constant (cm-1) of I-th internal rotation; E(K)=BINROT*K*K.
      DO 20 I=1,NINROT
   20 ROTINT=ROTINT*BINROT(I)
```

```
        ROTINT=SQRT(3.141593E0**NINROT/ROTINT)
     21 ROTCNT=ROTINT
        IF (NEXROT.LE.0) GO TO 23
        READ(5,#) BZ,BX
CARD#6  (If NEXROT > 0) BZ,BX: external rotational constant along
C        and perpendicular to axis of symmetric top, respectively.
        IF (BZ.GT.0.0E0) ROTCNT=ROTCNT*SQRT(3.141593E0/BZ)
        IF (BX.GT.0.0E0) ROTCNT=ROTCNT/BX
     23 CONTINUE
C*** Print rotational information here.
        READ(5,#) (IG(I),WE(I),XE(I),I=1,NOSC)
CARD#7  (IG(I),WE(I),XE(I),I=1,NOSC)
C        IG(I) = degeneracy of I-th oscillator;
C        WE(I) = harmonic vibrational frequency V of I-th oscillator
C                in cm-1;
C        XE(I) = anharmonicity factor X of I-th oscillator:
C                see [78.Y2]; here XE(I) << 1;
C                V = (1-X)*WO, where E(V)-E(0) = WO*V - WEXE*V**2.
C    OR  XE(I) = dissociation energy for oscillator/
C                activation energy, see [80.P1];
C                here XE(I) = 1 or XE(I) > 1.
C     compute energy levels for Morse oscillators:
        DO 40 I=1,NOSC
        G(I)=FLOAT(IG(I))
        GWE(I)=G(I)*WE(I)
        IF (XE(I).GE.1.0E0) GO TO 26
        IF (NMORPF.EQ.0) XE(I)=0.0E0
        XE(I)=XE(I)/(1.0E0+XE(I))
        WE(I)=WE(I)/(1.0E0-2.0E0*XE(I))
        WEXE(I)=WE(I)*XE(I)
        WEXE2(I)=2.0E0*WEXE(I)
        WO=WE(I)-WEXE(I)
        GO TO 27
     26 Y=4.0E0*ECRIT*XE(I)
        WO=Y/2.0E0*(1.0E0-SQRT(1.0E0-4.0E0*WE(I)/Y))
        WEXE(I)=WO*WO/Y
        WE(I)=WO+WEXE(I)
        XE(I)=WEXE(I)/WE(I)
C     Note: these two methods give slightly different results.
     27 CONTINUE
C     oscillator treated as harmonic if XE.LE.0.005:
        IF (XE(I).LE.5.0E-3) GO TO 40
        NMAX(I)=0.5E0*(1.0E0/XE(I)-1.0E0)+1.0E-4
        NMAX1=NMAX(I)+1
        E(I,1)=0.0E0
        DO 30 J=2,NMAX1
        FJ=FLOAT(J-1)
     30 E(I,J)=(WO-FJ*WEXE(I))*FJ
C*** Print vibrational information here.
     40 CONTINUE
        READ(5,#) (PCONC(I),I=1,NCONC)
CARD#8  (PCONC(I),I=1,NCONC); PCONC(I) is I-th pressure point (Torr).
        READ(5,#) EMATCH,EROUND,EMAX,ESTEP
CARD#9  EMATCH: maximum energy for direct count of state-sum (cm-1);
C        program will estimate (using steepest descent) sufficiency of
C        storage space for direct count up to the proposed value, and
C        (1) perform COUNT up to EMATCH if space sufficient,
C        (2) perform  counting up to a lower energy otherwise, or
C        (3) stop execution if estimated storage space greatly
C            exceeds that allocated.
```

```
C          EROUND: grain size for energy in direct count (cm-1);
C          EMAX: maximum energy for steepest descent calculation
C             of state sum: EMAX  > EMATCH;
C           ESTEP: energy grid size in steepest descent calculation.
        IF (EMATCH/EROUND+2.0E0.GT.FDIM) EROUND=EMATCH/(FDIM-2.0E0)
        NSTEP=ESTEP/EROUND+0.5E0
        ESTEP=EROUND*FLOAT(NSTEP)
        NPOINT=EMAX/ESTEP+1.5E0
        IF (NPOINT-NDIM2) 43,43,41
     41 ESTEP=EMAX/FLOAT(NDIM2-2)
        NSTEP=ESTEP/EROUND+0.001E0
        IF (EROUND*NSTEP.EQ.ESTEP) GO TO 42
        ESTEP=EROUND*(NSTEP+1)
        NSTEP=NSTEP+1
     42 NPOINT=EMAX/ESTEP+1.5E0
     43 CONTINUE
C    estimate sufficiency of storage space for direct count up to
C    EMATCH using steepest descent approximation.
        CALL DESENT (1,0,1.0E0,NOSC,NMORPF,EMATCH,1,0,ESTEP)
        SUMEST=P(1)
        IF (SUMEST.LE.0.9E0*FDIM) GO TO 51
        IF (SUMEST-1.5E0*FDIM) 46,46,44
     44 CONTINUE
C*** Print flag that storage is insufficient, and stop.
        GO TO 999
     46 NMATCH=EMATCH*FDIM/(SUMEST*EROUND)
        EMATCH=EROUND*FLOAT(NMATCH)
C*** Print flag that EMATCH was lowered by program.
     51 CONTINUE
C*** Print out graining information.
        NGRID=EMATCH/EROUND+1.5E0
C    direct count of state-sum up to EMATCH:
        CALL COUNT (EMATCH,EROUND,NOSC,ICOUNT,NGRID)
        NSTORE=(NGRID-1)/NSTEP*NSTEP+1
        IF (NEXROT) 55,55,61
     55 CONTINUE
        DO 60 I=2,NGRID
     60 P(I)=P(I)+P(I-1)
C    convolute with internal rotational states, if necessary:
        IF (NINROT.GT.0) CALL RINCNT (EMATCH,EROUND,NGRID,NINROT)
        GO TO 102
C    convolute with external rotational states, if necessary:
     61 CALL EXTROT (EMATCH,EROUND,NGRID,NEXROT,BZ,BX)
        DO 80 I=2,NGRID
     80 P(I)=P(I)+P(I-1)
        IF (NINROT.GT.0) CALL RINCNT (EMATCH,EROUND,NGRID,NINROT)
        DO 100 I=1,NSTORE,NSTEP
           NI=NSTORE-I+1
           DO 90 J=2,ICOUNT
              NIJ=NI+1-IE(J)
              IF (NIJ.LE.0) GO TO 100
     90    P(NI)=P(NI)+P(NIJ)*ELEVEL(J)
    100 CONTINUE
    102 NSTORE=(NSTORE-1)/NSTEP+1
        NSUBTR=1-NSTEP
        DO 120 I=2,NSTORE
    120 P(I)=P(NSTEP*I+NSUBTR)
        EMIN=FLOAT(NSTORE)*ESTEP
        NRUN=(EMAX-EMIN)/ESTEP+1.5E0
C    compute state-sum below reaction threshold by steepest descent:
```

```
        CALL DESENT (1,NROT,ROTCNT,NOSC,NMORPF,EMIN,NRUN,NSTORE,ESTEP)
        NPOINT=NRUN+NSTORE
C      compute state-density for reactive states by steepest descent:
        CALL DESENT (0,NROT,ROTCNT,NOSC,NMORPF,ECRIT,NPOINT,NDIM2,ESTEP)
C*** Print state-sum and state-density if desired:
   141 IF (NPRINT.EQ.0) GO TO 201
        N5=NPRINT*5
        NPRT=NPOINT/N5
        DO 180 IPRINT=1,NPRT
            DO 160 J=1,5
                JINDEX=NPRT*(J-1)+IPRINT
                EPRINT(J)=FLOAT(JINDEX-1)*ESTEP
                PPRINT(J)=P(JINDEX)
   160      CONTINUE
C*** Print sum of states in five columns.
   180 CONTINUE
        DO 200 IPRINT=1,NPRT
            DO 190 J=1,5
                JINDEX=NPRT*(J-1)+IPRINT
                EPRINT(J)=FLOAT(JINDEX-1)*ESTEP+ECRIT
                PPRINT(J)=P(NDIM2+JINDEX)
   190      CONTINUE
C*** Print density of states in five columns.
   200 CONTINUE
   201 CONTINUE
        READ(5,#) TEMPK,RELINT,RANDK1,RANDK2,WCRIT,XSECTN,ZMA,ZM3RD
CARD#10   TEMPK: temperature of reaction (deg K); TEMPK=0 - stop.
C         RELINT: collisional energy relaxation rate (/sec/Torr);
C         RANDK1: intrinsic randomisation rate (sec-1);
C         RANDK2: collisional randomisation rate (/sec/Torr);
C         WCRIT: frequency of critical oscillator (cm-1);
C         If RANDK=0, only strong collision calculation is executed;
C         If RANDK>0, exact AND separable calculations are done.
C         If RELINT is unavailable, use:
C         XSECTN: collisional diameter (A), in position 6;
C         ZMA: mass of reactant (gm/mole), in position 7;
C         ZM3RD: mass of third body (gm/mole), in position 8.
        IF (TEMPK.LE.0.0E0) GO TO 999
C      collisional relaxation rate by either input data or
C      computation by hard sphere model:
        IF (RELINT.LE.0.0E0)RELINT=XSECTN**2*SQRT((ZMA+ZM3RD)/
     1                            (TEMPK*ZMA*ZM3RD))*4.414518E7
        BETA=1.438833E0/TEMPK
        RINF=AFACTR*EXP(-BETA*ECRIT)
        WCRIT=6.0E10*WCRIT
C*** Print all kinetic parameters here.
        ROTOR=0.5E0*FLOAT(NROT)
        CALL PFNCTN (BETA,PF,ROTCNT,ROTOR,NOSC)
C      coarse graining of reactive states:
        CALL GRAIN (NCHANL,NROT,BETA,ECRIT,AFACTR,PF,ESTEP,ICOUNT,EROUND,
     1 RELINT,NDIM2,NPOINT)
C      computation of fall-off rate expression:
        CALL UNIRAT (NCHANL,RELINT,NCONC)
        GO TO 201
   999 CONTINUE
        STOP
        END
```

```
      SUBROUTINE UNIRAT (N,ZCOLL,NCONC)
C     UNIRAT calculates the unimolecular rate constant at a series
C     of pressures by methods described in chapters 5 and 7;
C     maximum of 40 channels and 100 pressures.
C     COMMON BLOCK 12 provides means of calculating k/k(infinity).
      REAL*4 E(40),P(40),R(40),RR(40),PCONC(100),UNIK(100)
      REAL*4 SEPRAT(100),RANRAT(100)
      REAL*8 DRELMU,DWCRIT,DDR,DZERO,DONE
      REAL*8 COR,TT,RC1,RC3,PNM,PDM,COF,X,PHI
      REAL*8 BET(40),BETO(40),BET1(40),CAP(2)
      COMMON/BLK9/PCONC,UNIK
      COMMON/BLK11/E,P,R
      COMMON/BLK12/RINF,RANDK1,RANDK2,WCRIT
      ZERO=0.0E0
      ONE=1.0E0
      DZERO=0.0D0
      DONE=1.0D0
      DWCRIT=WCRIT
C     for consistency checks and for Figures 5.3, 5.4, and 5.5:
C        SUMPI=ZERO
C        SUMPR=ZERO
C        DO 201 I=1,N
C           PR=P(I)*R(I)
C           PLG=ALOG10(P(I))
C           RLG=ALOG10(R(I))
C           SUMPI=SUMPI+P(I)
C           SUMPR=SUMPR+PR
C 201 CONTINUE
      DO 300 J=1,NCONC
         RELMU=PCONC(J)*ZCOLL
         SUMNUM=ZERO
         DO 210 I=1,N
            SUMNUM=SUMNUM+P(I)*R(I)/(RELMU+R(I))
  210    CONTINUE
C     choice of lower bound, equation (5.14):
         UNIK(J)=RELMU*SUMNUM
C     or upper bound, equation (5.17):
C        UNIK(J)=RELMU*SUMNUM/(ONE-SUMNUM)
C     or else iterate to exact solution [81.N;81.V2].
C     strong collision calculation ends here.
         DR=RANDK1+PCONC(J)*RANDK2
C     variable randomisation rates can be included here: use DR(I).
         IF (DR.LE.ZERO) GO TO 209
         DR=DR+RELMU
C     separable approximation, see [80.P1]:
         SUMNUM=ZERO
         DO 1 I=1,N
            RR(I)=DR*R(I)/(DR+WCRIT-R(I))
            SUMNUM=SUMNUM+P(I)*RR(I)/(RELMU+RR(I))
    1    CONTINUE
         SEPRAT(J)=RELMU*SUMNUM/(ONE-SUMNUM)
C     equation (7.7):
         DDR=DR
         DRELMU=RELMU
         PHI=DZERO
         DO 211 I=1,N
            BET(I)=P(I)
            BET1(I)=R(I)*P(I)/WCRIT
```

```
          BETO(I)=BET(I)-BET1(I)
          TT=(DDR+DWCRIT)/2.0D0
          COR=TT*TT - BET1(I)*DDR*DWCRIT/BET(I)
          COR=DSQRT(COR)
          CAP(1)=TT-COR
          CAP(2)=TT+COR
          DO 20 K=1,2
              RC1=DDR-CAP(K)
              RC3=DONE + DWCRIT/RC1
              PNM=BETO(I)+BET1(I)/RC3
              PNM=PNM*PNM
              PDM=BETO(I)+BET1(I)/RC3/RC3
              COF=PNM/PDM
              X=CAP(K)
              PHI=PHI+COF*X/(DONE + X/DRELMU)
   20         CONTINUE
  211     CONTINUE
          RANRAT(J)=PHI/(DONE-PHI/DRELMU)
  209     CONTINUE
C*** Print required pressure and rate data here.
  300 CONTINUE
      RETURN
      END
```

```
C        UTILITY PACKAGE
C        unless otherwise stated, all variables in subroutine
C        parameter lists and common blocks are defined at point
C        of calling and are left unchanged upon return.

         SUBROUTINE COUNT (EMAX,EROUND,NOSC,ICOUNT,NSTORE)

C    COUNT performs direct count of the state-degeneracy for a
C    set of independent Morse/harmonic oscillators, and returns
C    (1)    P(I) = state degeneracy at E = EROUND*(I-1); and
C    (2)    (ELEVEL(I), IE(I), I=1, ICOUNT);
C    where (IE(I)-1)*EROUND = energy of I-th state, and
C    ELEVEL(I) = state-degeneracy of the I-th state.

         REAL*4 VMU(25),VMUX(25),VXE(25),P(4000),ELEVEL(4000)
         INTEGER IG(25),IE(4000)
         COMMON/BLK1/P
         COMMON/BLK2/ELEVEL,IE
         COMMON/BLK5/VMU,IG
         COMMON/BLK6/VXE
         COMMON/BLK8/VMUX
C    initialisation of state-degeneracy array:
         P(1)=1.0E0
         IE(1)=1
         MCUR=1
         IDRIVE=0
         DO 10 I=2,NSTORE
      10 P(I)=0.0E0
         DO 120 IOSC=1,NOSC
         NI=NOSC-IOSC+1
         E=VMU(NI)-2.0E0*VMUX(NI)
         IF (E.GT.EMAX) GO TO 120
         WE=VMU(NI)
         WEXE=VMUX(NI)
         WO=WE-WEXE
         NLEVEL=EMAX/WE
         IF (VXE(NI).GT.5.0E-3) NLEVEL=0.5E0*(1.0E0/VXE(NI)-1.0E0)
         NDEG=IG(NI)
         DO 100 IDEG=1,NDEG
C    locate energy levels of first oscillator:
         IF (IDRIVE) 11,11,21
      11       IDRIVE=1
             DO 20 I=1,NLEVEL
                 MCUR=MCUR+1
                 E=(WO-I*WEXE)*I
                 IF (E.GT.EMAX) GO TO 100
                 M=E/EROUND+1.5E0
                 IE(MCUR)=M
                 P(M)=1.0E0
      20       CONTINUE
             MCUR=MCUR+1
             GO TO 100
      21       CONTINUE
             DO 60 I=1,NLEVEL
                 E=(WO-I*WEXE)*I
                 IF (E.GT.EMAX) GO TO 100
                 MADD=E/EROUND+0.5E0
C    convolution of energy levels of I-th oscillator with
C    those of the first (I-1) oscillators:
                 DO 40 M=1,MSTORE
```

```
                IEM=IE(M)
                IEMADD=IEM+MADD
                IF (IEMADD.GT.NSTORE) GO TO 40
                IE(MCUR)=IEMADD
                MCUR=MCUR+1
                P(IEMADD)=P(IEMADD)+1.0E0
   40           CONTINUE
   60        CONTINUE
  100   MSTORE=MCUR-1
  120 CONTINUE
      ICOUNT=1
      ELEVEL(1)=1.0E0
      IE(1)=1
C     generate IE and ELEVEL arrays:
      DO 140 I=2,NSTORE
         IF (P(I).EQ.0.0E0) GO TO 140
         ICOUNT=ICOUNT+1
         IE(ICOUNT)=I
         ELEVEL(ICOUNT)=P(I)
  140 CONTINUE
      RETURN
      END
```

```
      SUBROUTINE EXTROT (EMAX,EROUND,NSTORE,NEXROT,BZ,BX)

C     EXTROT performs direct count of external rotational states
C     of a symmetric top, given the rotational constant(s).
C     Upon return, P(I) = state-degeneracy at E = EROUND*(I-1);
C     it need not be initialised beforehand.

      REAL*4 P(4000)
      COMMON/BLK1/P
      P(1)=1.0E0
      DO 10 I=2,NSTORE
   10 P(I)=0.0E0
C     NEXROT =1 --> Z-axis rotation only; E(K) = BZ*K*K;
C            =2 --> XY-plane rotation only: E(J) = BX*J*(J+1);
C            =3 --> symmetric or spherical top rotation;
C                   E(J,K) = BX*J(J+1) + (BZ-BX)*K*K.
C     see DAVIDSON [62.D] or PITZER [53.P1] for relevant formulae.
      GO TO (1001,2001,3001), NEXROT
 1001 KMAX=SQRT(EMAX/BZ)
      DO 20 K=1,KMAX
         BK=BZ*FLOAT(K*K)
         MK=BK/EROUND+1.5E0
         P(MK)=P(MK)+2.0E0
   20 CONTINUE
      GO TO 4001
 2001 JMAX=0.5E0*(SQRT(1.0E0+4.0E0*EMAX/BX)-1.0E0)
      DO 40 J=1,JMAX
         BJ=BX*FLOAT(J*J+J)
         MJ=BJ/EROUND+1.5E0
         IF (MJ.LE.NSTORE) P(MJ)=P(MJ)+2.0E0*FLOAT(J)+1.0E0
   40 CONTINUE
      GO TO 4001
 3001 BZMBX=BZ-BX
      IF (BZMBX) 51,101,201
   51 NMAXJ=(-BX+SQRT(BX*BX+4.0E0*BZ*EMAX))/(2.0E0*BZ)
      DO 60 J=1,NMAXJ
         BJ=BX*FLOAT(J*J+J)
         MJ=BJ/EROUND+1.5E0
         IF (MJ.LE.NSTORE) P(MJ)=2.0E0*J+1.0E0+P(MJ)
   60 CONTINUE
      DO 80 J=1,NMAXJ
         BJ=BX*FLOAT(J*J+J)
         DO 70 K=1,J
         JK=J-K+1
         MJK=(BJ+BZMBX*FLOAT(JK*JK))/EROUND+1.5E0
         IF (MJK.GT.NSTORE) GO TO 80
   70    P(MJK)=P(MJK)+4.0E0*FLOAT(J)+2.0E0
   80 CONTINUE
      GO TO 4001
  101 NMAXJ=SQRT(EMAX/BX)
      DO 120 J=1,NMAXJ
         MJ=BX*FLOAT(J*J+J)/EROUND+1.5E0
         IF (MJ.GT.NSTORE) GO TO 4001
         P(MJ)=P(MJ)+(2.0E0*FLOAT(J)+1.0E0)**2
  120 CONTINUE
      GO TO 4001
  201 NMAXJ=SQRT(EMAX/BX)
      DO 220 J=1,NMAXJ
         MJ=BX*FLOAT(J*J+J)/EROUND+1.5E0
         IF (MJ.GT.NSTORE) GO TO 221
```

```
          P(MJ)=2.0E0*FLOAT(J)+1.0E0+P(MJ)
  220 CONTINUE
  221 CONTINUE
      DO 260 J=1,NMAXJ
          BJ=BX*FLOAT(J*J+J)
          DO 240 K=1,J
              MJK=(BJ+BZMBX*FLOAT(K*K))/EROUND+1.5E0
              IF (MJK.GT.NSTORE) GO TO 260
  240         P(MJK)=P(MJK)+4.0E0*FLOAT(J)+2.0E0
  260 CONTINUE
 4001 CONTINUE
      RETURN
      END




      SUBROUTINE RINCNT (EMAX,EROUND,NPOINT,NINROT)

C     RINCNT performs direct count of 1-dimensional rotational levels
C     up to EMAX and convolutes them with the state-sum of other
C     degrees of freedom.
C     P(I) is state-sum of other degrees of freedom at I-th energy grid
C     point upon entry, and is the convoluted state-sum upon return.

      REAL*4 P(4000),BINROT(5)
      INTEGER IE(1000)
      COMMON/BLK1/P
      COMMON/BLK10/BINROT
      DO 100 I=1,NINROT
          BCNST=BINROT(I)
          NLEVEL=SQRT(EMAX/BCNST)
C     generate 1-dimensional internal rotational energy levels:
          DO 50 J=1,NLEVEL
   50         IE(J)=BCNST*FLOAT(J*J)/EROUND+0.5E0
C     convolution:
          DO 70 K=2,NPOINT
              NK=NPOINT-K+1
              DO 60 J=1,NLEVEL
                  NKIJ=NK-IE(J)
                  IF (NKIJ.LE.0) GO TO 70
   60             P(NK)=P(NK)+2.0E0*P(NKIJ)
   70         CONTINUE
  100 CONTINUE
      RETURN
      END
```

```
      SUBROUTINE FNPHI1 (BETA,PF,SUM1,SUM2,SUMO,NOSC,ROTOR)

C     FNPHI1 computes the reduced partition function PF, the average
C     energy SUMO and its first two derivatives SUM1 and SUM2 with
C     respect to the inverse temperature BETA at a given BETA, for a
C     set of NOSC Morse oscillators and ROTOR 1-dimensional rotors.
C     Exact partition function is used for any Morse oscillator
C     having anharmonicity constant > 0.005;
C     rotors are treated as classical.

      REAL*4 E(25,100),VMU(25),VMUX(25),GVMU(25),G(25),VXE(25)
      INTEGER IG(25),NMAX(25)
      COMMON/BLK3/G,GVMU
      COMMON/BLK4/E,NMAX
      COMMON/BLK5/VMU,IG
      COMMON/BLK6/VXE
      COMMON/BLK8/VMUX
      BOLT=1.0E0/BETA
      SUMO=0.0E0
      SUM1=0.0E0
      SUM2=0.0E0
C     reduced rotational partition function:
      PROD=BOLT**ROTOR
      DO 50 I=1,NOSC
      IF (VXE(I)-5.0E-3) 21,21,31
C     calculate quantities for harmonic oscillators; Morse oscillators
C     with anharmonicity factor less than 0.005 treated as harmonic:
   21 EXPINV=EXP(-VMU(I)*BETA)
      PFHO=1.0E0/(1.0E0-EXPINV)
      PROD=PROD*PFHO**IG(I)
      EHO=GVMU(I)*PFHO*EXPINV
      SUMO=SUMO+EHO
      EHO1=-VMU(I)*PFHO*EHO
      EHO2=-VMU(I)*PFHO*EHO1*(1.0E0+EXPINV)
      SUM1=SUM1+EHO1
      SUM2=SUM2+EHO2
      GO TO 50
C     calculate quantities for Morse oscillators:
   31 CONTINUE
      NMAX1=NMAX(I)+1
      PFMO=1.0E0
      EMO=0.0E0
      EMO1=0.0E0
      EMO2=0.0E0
      DO 40 J=2,NMAX1
      TERM=-BETA*E(I,J)
      IF (TERM.LT.-1.0E2) GO TO 41
      DTERM=EXP(TERM)
      PFMO=PFMO+DTERM
      EMMO=EMO-E(I,J)*DTERM
      EMO1=EMO1+E(I,J)*E(I,J)*DTERM
      EMO2=EMO2-E(I,J)**3*DTERM
   40 CONTINUE
   41 PROD=PROD*PFMO**IG(I)
      EMOR=-EMO/FFMO
      EMOR1=EMOR*EMOR-EMO1/PFMO
      EMOR2=2.0E0*EMOR*EMOR1-EMO2/PFMO-EMOR*EMO1/PFMO
      SUMO=SUMO+EMOR*G(I)
      SUM1=SUM1+EMOR1*G(I)
      SUM2=SUM2+EMOR2*G(I)
```

```
 50 CONTINUE
    PF=PROD
    SUM0=SUM0+ROTOR*BOLT
    SUM1=SUM1-ROTOR*BOLT*BOLT
    SUM2=SUM2+2.0E0*ROTOR*BOLT**3
    RETURN
    END




    SUBROUTINE PFNCTN (BETA,PF,ROTCNT,ROTOR,NOSC)
C   PFNCTN calculates the partition function PF of the system at
C   inverse temperature BETA.

    REAL*4 WE(25),XE(25),E(25,100)
    INTEGER IG(25),NMAX(25)
    COMMON/BLK4/E,NMAX
    COMMON/BLK5/WE,IG
    COMMON/BLK6/XE
    BOLT=1.0E0/BETA
C   rotational partition function:
    PROD=BOLT**ROTOR*ROTCNT
C   vibrational partition function (exact for Morse oscillators):
    DO 50 I=1,NOSC
        IF (XE(I).GE.5.0E-3) GO TO 21
 11     EXPV=EXP(WE(I)*BETA)
        FRAC=EXPV/(EXPV-1.0E0)
        GO TO 31
 21     FRAC=1.0E0
        NMAX1=NMAX(I)+1
        DO 30 J=2,NMAX1
            TERM=-BETA*E(I,J)
            IF (TERM.LT.-2.0E1) GO TO 31
            FRAC=FRAC+EXP(TERM)
 30     CONTINUE
 31     PROD=PROD*FRAC**IG(I)
 50 CONTINUE
    PF=PROD
    RETURN
    END
```

```
      SUBROUTINE FNPHI2 (BETA,PF,SUM1,SUM2,SUMO,NOSC,ROTOR,MORSE)

C     FNPHI2 computes and returns the reduced partition function PF,
C     the average internal energy SUMO and its first two derivatives
C     SUM1 and SUM2 with respect to the inverse temperature parameter
C     BETA for a set of Morse oscillators and 1-dimensional rotors,
C     using the HOARE-RUIJGROK approximation [70.H2].

      REAL*4 VMU(25),PFHO(25),EHO(25),EHO1(25),EHO2(25),G(25)
      REAL*4 GVMU(25),VMUXE2(25)
      INTEGER IG(25)
      COMMON/BLK3/G,GVMU
      COMMON/BLK5/VMU,IG
      COMMON/BLK7/VMUXE2
      BOLT=1.0E0/BETA
      SUMO=0.0E0
      SUM1=0.0E0
      SUM2=0.0E0
C     reduced rotational partition function:
      PROD=BOLT**ROTOR
C     for harmonic oscillators:
      DO 20 I=1,NOSC
          EXPINV=EXP(-BETA*VMU(I))
          PFHO(I)=1.0E0/(1.0E0-EXPINV)
          PROD=PROD*PFHO(I)**IG(I)
          EHO(I)=GVMU(I)*PFHO(I)*EXPINV
          EHO1(I)=-VMU(I)*PFHO(I)*EHO(I)
          EHO2(I)=-VMU(I)*PFHO(I)*EHO1(I)*(1.0E0+EXPINV)
          SUMO=SUMO+EHO(I)
          SUM1=SUM1+EHO1(I)
          SUM2=SUM2+EHO2(I)
   20 CONTINUE
      IF (MORSE.EQ.0) GO TO 41
C     anharmonic corrections:
      DO 40 I=1,NOSC
          IF (VMUXE2(I).EQ.0.0E0) GO TO 40
          FCORR0=PFHO(I)**2*BETA*EXP(-VMU(I)*BETA)
          PROD=PROD*(1.0E0+VMUXE2(I)*FCORR0)**IG(I)
          CNST1=BOLT-VMU(I)-2.0E0*EHO(I)
          FCORR1=FCORR0*CNST1
          CNST2=-(2.0E0*EHO1(I)+BOLT**2)
          FCORR2=FCORR1*CNST1+FCORR0*CNST2
          CNST3=2.0E0*(BOLT**3-EHO2(I))
          FCORR3=2.0E0*FCORR1*CNST2+FCORR2*CNST1+FCORR0*CNST3
          SUMO=SUMO-VMUXE2(I)*FCORR1*G(I)
          SUM1=SUM1-VMUXE2(I)*FCORR2*G(I)
          SUM2=SUM2-VMUXE2(I)*FCORR3*G(I)
   40 CONTINUE
   41 PF=PROD
      SUMO=SUMO+ROTOR*BOLT
      SUM1=SUM1-ROTOR*BOLT*BOLT
      SUM2=SUM2+ROTOR*2.0E0*BOLT**3
      RETURN
      END
```

```
      SUBROUTINE DESENT (NFLAG,NROT,ROTCNT,NOSC,MORSE,EMIN,NRUN,NADD,
     1 ESTEP)
C     DESENT computes the density or sum of states using the first order
C     steepest descent method, and returns same via P(I). Algorithm
C     based on the non-iterative procedure of YAU and PRITCHARD [77.Y2].
C     It gives the state-sum or state-density at NRUN energy values
C     starting from EMIN at an increment of ESTEP; the result
C     for the I-th energy value is returned via P(I+NADD).

      REAL*4 G(25),GVMU(25),P(4000)
      COMMON/BLK1/P
      COMMON/BLK3/G,GVMU
C     FLAG:     NFLAG=0 for state-density, =1 for state-sum.
      ROTOR=0.5E0*FLOAT(NROT+2*NFLAG)
      REDFAC=3.989423E-01*ROTCNT
      ENERGY=EMIN
      EMIN=EMIN-ESTEP
      EINV=1.0E0/ENERGY
C     initialisation of approximation to beta:
      SUM1=G(1)
      SUMVMU=GVMU(1)
      DO 15 I=2,NOSC
         IF (SUMVMU.GT.ENERGY) GO TO 16
         SUM1=SUM1+G(I)
         SUMVMU=SUMVMU+GVMU(I)
   15 CONTINUE
   16 VMUINV=SUM1/SUMVMU
      BETA0=ROTOR*EINV+VMUINV*ALOG(1.0E0+SUMVMU*EINV)
C     oscillator type:   =0 for harmonic oscillator,
C                        =1 for Morse oscillator using exact quantities,
C                        =2 for Morse oscillator using approx. formulae.
      IF (MORSE-1) 18,21,18
   18 CALL FNPHI2 (BETA0,PF,DEO,DDEO,ESUM,NOSC,ROTOR,MORSE)
      GO TO 22
   21 CALL FNPHI1 (BETA0,PF,DEO,DDEO,ESUM,NOSC,ROTOR)
   22 CONTINUE
C     non-iterative procedure begins:
      DO 100 IRUN=1,NRUN
         ENERGY=ESTEP*FLOAT(IRUN)+EMIN
         DEOINV=1.0E0/DEO
         DELTAE=ENERGY-ESUM
      BETA=BETA0+DELTAE*DEOINV/(1.0E0+0.5E0*DELTAE*DEOINV*DEOINV*DDEO)
         IF (MORSE-1) 28,31,28
   28    CALL FNPHI2 (BETA,PF,DEO,DDEO,ESUM,NOSC,ROTOR,MORSE)
         GO TO 32
   31    CALL FNPHI1 (BETA,PF,DEO,DDEO,ESUM,NOSC,ROTOR)
   32    CONTINUE
         DENSTY=REDFAC*EXP(BETA*ENERGY)*PF/SQRT(-DEO)
         P(IRUN+NADD)=DENSTY
         BETA0=BETA
  100 CONTINUE
      EMIN=EMIN+ESTEP
      RETURN
      END
```

```
      SUBROUTINE GRAIN (NCHANL,NROT,BETA,ECRIT,AFACTR,PF,ESTEP,NLEVEL,
     1 EROUND,RELINT,NDIM2,NPOINT)
C     GRAIN groups the states in the reactive energy range (from
C     ECRIT to ECRIT + EMAX) into NCHANL pseudolevels, and calculates
C     the microscopic rate constants for the pseudolevels by using the
C     inverse LAPLACE transform relation for the ARRHENIUS equation.
C     Upon return, EGRAIN(I) = energy of I-th pseudolevel;
C                  POPLTN(I) = fractional population of I-th pseudolevel;
C                  RCNST(I)  = decay rate constant of I-th pseudolevel.

      REAL*4 EGRAIN(40),RCNST(40),POPLTN(40),P(4000),ELEVEL(4000)
      INTEGER IE(4000)
      COMMON/BLK1/P
      COMMON/BLK2/ELEVEL,IE
      COMMON/BLK11/EGRAIN,POPLTN,RCNST
      CNST=0.5E0*(1.0E0+EXP(BETA*ESTEP))
      NPNT1=NPOINT-1
    1 NGRID=NPNT1/NCHANL
      NSTART=1
      DO 100 I=1,NCHANL
         ISTART=NGRID*(I-1)+1
         IEND=NGRID*I
         SUMP=0.0E0
         SUMR=0.0E0
         EBEGIN=ESTEP*FLOAT(ISTART-1)
         EGRAIN(I)=ESTEP*NGRID*(I-1)+ECRIT
         EEND=ESTEP*FLOAT(IEND-1)
C     generate rate constant using inverse LAPLACE transform relation:
      IF (NROT) 11,11,41
   11    ELEV=EROUND*FLOAT(IE(NLEVEL)-1)
         IF (EEND.GT.ELEV) GO TO 41
   21    DO 30 J=NSTART,NLEVEL
            EJ=EROUND*FLOAT(IE(J)-1)
            IF (EJ.LT.EBEGIN) GO TO 30
            IF (EJ.GE.EEND) GO TO 31
            SUMR=SUMR+ELEVEL(J)*EXP(-BETA*EJ)
   30    CONTINUE
   31    SUMR=SUMR*EXP(-BETA*ECRIT)/PF
         NSTART=J-1
         GO TO 71
   41    DO 60 J=ISTART,IEND
            TERM=EXP(-BETA*FLOAT(J)*ESTEP)
            SUMR=SUMR+(P(J+1)-P(J))*TERM
   60    CONTINUE
         SUMR=SUMR*CNST/PF/EXP(BETA*ECRIT)
C     graining of energy levels:
   71    DO 80 J=ISTART,IEND
            TERM=-BETA*ESTEP*FLOAT(J-1)
            SUMP=SUMP+P(NDIM2+J)*EXP(TERM)
   80    CONTINUE
         SUMP=SUMP-0.5E0*(P(NDIM2+ISTART)*EXP(-BETA*EBEGIN)
     1                    +P(NDIM2+IEND)*EXP(-BETA*EEND))
         SUMP=SUMP*ESTEP/PF/EXP(BETA*ECRIT)
         SUMR=SUMR*AFACTR/SUMP
         RCNST(I)=SUMR
         POPLTN(I)=SUMP
         IF (SUMR.LE.0.0E0) GO TO 121
  100 CONTINUE
      RETURN
```

```
C      adjust grain size, if a pseudolevel is non-reactive:
  121 NCHANL=NCHANL*3/4
C*** Print message that grain size is too fine and that
C      fewer channels have been tried.
      GO TO 1
      END




CH3NC FALL-OFF: DATA OF SCHNEIDER AND RABINOVITCH
  4.295   E 13 3.84    E 01
    0  70  12
    8    0    3    2
5.22       0.33257
  2 263.       0.020   1 945.0      0.020    2 1129.0      0.015
  1 1429.0     0.025   2 1467.0     0.025    1 2166.0      0.020
  1 2966.0     0.080   2 3014.0     0.040
  1.0E-02 2.0E-02 3.0E-02 .......
                  70 values of the pressure
                        ....... 9.0E 03 1.0E 04 1.0E 05
2100.      1.0       6000.      15.
503.55 1.173E+06 4.0  E+12 1.25 E+12 263.0
0.00




C3H6 FALL-OFF: DATA OF FALCONER, HUNTER AND TROTMAN-DICKENSON
  1.88    E 15 6.56    E 01
    0  70  15
   14    0    3    2
0.435      0.670
  2 739.       0.015   1 854.       0.015    2 869.        0.018
  1 963.       0.018   2 1029.      0.020    1 1133.       0.020
  1 1188.      0.0     2 1188.      0.020    2 1442.       0.025
  1 1479.      0.0     2 3028.      0.053    1 3038.       0.080
  2 3082.      0.043   1 3103.      0.029
  1.0E-04 3.0E-04 5.0E-04 .......
                  70 values of the pressure
                        ....... 4.0E 03 5.0E 03 1.0E 04
1700.      1.0          12000.    25.
765.0  0.0  E 00 0.0  E 00 0.0  E 00 963.0  1.265E 00   42.0   42.0
0.00
```

# EXERCISES

The textbook by Forst [73.F] provides a number of interesting exercises which I do not propose to duplicate. Instead, I will suggest below a selection of computational experiments, all of which can be undertaken once the basic algorithms in Appendix 3 have been assembled together in an available machine and the inevitable organisational conflicts have been resolved. I would suggest that, initially, the student should practise with a few computer runs in which only the input parameters are varied, in order to find out the general behaviour of the method; several of these are listed first. Then follow a number of fairly extensive numerical experiments, each marked *, which are designed to illustrate relevant points of the theory; most students will not have the time to venture beyond these. Finally, for the student whose concern is primarily with the unimolecular kinetic problem, there are three marked ** and ***, which are more involved: the first two can be completed either by a small amount of reprogramming of the routines MAIN and UNIRAT, or by the construction of small routines specially designed for the problem, but the last one should not be attempted without a thorough understanding of the procedures used to calculate sums and densities of states.

Any reaction may be chosen as the subject for each calculation, but the methyl isocyanide isomerisation provides a fairly realistic example of a strong collision reaction without consuming too much computer time; data sets are given for both methyl isocyanide and cyclopropane.

1 Examine the convergence of the computed rate constant at a series of pressures as the number of reactive grains is increased, and confirm that about 10–15 equally spaced grains are adequate. By a separate (hand?) calculation show that if only one grain is used, the rate constant variation with pressure is strict Lindemann.

2 Examine the difference in the computed rate constants caused by using the harmonic or anharmonic oscillator models for the same reactant molecule; also, examine the differences found by using the approximate and the exact Morse partition functions, *but beware* that the use of the exact partition functions will be almost an order of magnitude more time consuming.

3 For your chosen reaction, examine the effects of varying, one at a time, the rotational constants, selected vibrational frequencies and their respective anharmonicities upon the density of states $\rho(E)$, the specific rate function $k(E)$, and the rate constant $k_{uni,p}$; also, examine the effect of omitting the external rotational constants altogether.

4 Examine what happens if the $A$ factor for the reaction is assumed to be a factor of 10 larger (or smaller) than normal.

5 For a strict Arrhenius strong collision reaction, examine whether the shape of the fall-off continues to get flatter as $T \to \infty$, or does it eventually become sharper again, as suggested by [65.P]; if so, why?

6 What kind of variation of $\mu$ with $T$ would be required to keep $p_{\frac{1}{2}}$ constant for your chosen reaction?

7 Show that the magnitude of the feature in Figure 7.3 becomes larger if the internal relaxation rate $\mu$ is made ten times larger, and that it almost disappears if it is made ten times smaller.

8 *Plot the curve shown in Figure 5.2 for an infinite sphere and for a 1 l sphere, both with $4 \times 10^{-4}$ Torr of water vapour present; the relative collision efficiency of water is 0.8 on a pressure for pressure basis.

9 *Investigate the difference between $E_\infty$ and $E_0$ as a function of molecular complexity at a given temperature, and as a function of temperature for a given molecule.

10 *Calculate the mean energy of the reacting molecules at various pressures; also calculate by equation (5.24) or (5.28) the activation energy for each pressure and investigate the transition between equations (5.26) and (5.27) as the pressure is changed. It may be helpful to have read [32.K; 78.T3].

11 *Compare the shape of the fall-off for a given reaction which is assumed to be either strict Arrhenius in form (i.e. the standard calculation), or modified Arrhenius in form; you will need to use equation (4.12) for the specific rate function in this case.

12 *Use equation (5.14) to evaluate the rate constant, but form the decay rate constants according to the Kassel prescription (1.7); show that this is the same result as that obtained by using the tabulated values of Kassel integrals [72.E] for the same value of $s$.

13 *Plot a series of five strict Lindemann curves, all with the same value of $k_0$, but with successive values of $k_\infty = 1$, $10^{\frac{1}{2}}$, $10$, $10^{\frac{3}{2}}$, and $100$. Examine the resulting fall-off curve formed by adding these five basic curves with various weight factors.

14 *Construct a simple extended Lindemann expression, for example

$$k_{\text{uni}} = \frac{1}{2}\left(\frac{ap}{1+bp} + \frac{sap}{1+sbp}\right) \quad \text{or} \quad \frac{1}{2}\left(\frac{ap}{1+bp} + \frac{ap}{1+sbp}\right)$$
$$(s>1) \qquad\qquad\qquad (s<1)$$

and compare the resulting fall-off shapes with those found in two or three real cases; is there any correspondence between the values of $s$ required in these equations and the appropriate value of the Kassel $s$ parameter? If you use the second of these two suggested forms, you may find a rather feeble conjecture, equation 7 of [77.P], useful.

15 **By using equation (12) of [81.V3], examine how the shape of the fall-off changes as the bottleneck is moved in either direction away from the threshold level. Do the results make physical sense?

16 **Rewrite the routine UNIRAT to accommodate two (or more) competing strong collision reactions having different thresholds, and test it out on the various reactions of monofluorocyclopropane [64.C; 78.F1]; compare your results with Figure 5.10. (Notice that it will be necessary to choose a grain width which is smaller than the difference in energy of the respective thresholds, so that considerably more than the usual 10 or 15 strips will be required if they are all of equal energy spacing.)

17 ***Rewrite the appropriate parts of the UTILITY routines to incorporate the much more efficient Laguerre quadrature [64.H; 67.H], rather than the strip integration which is now used to calculate the rate constant.

# REFERENCES

## Before 1950

[20.L]  I. Langmuir. *J. Amer. Chem. Soc.*, **42**, 2190–205 (1920).
[20.T]  R. C. Tolman. *J. Amer. Chem. Soc.*, **42**, 2506–28 (1920).
[22.L]  F. A. Lindemann. *Trans. Faraday Soc.*, **17**, 598–9 (1922).
[23.C]  J. A. Christiansen & H. A. Kramers. *Z. physikalische Chem.*, **104**, 451–71 (1923).
[27.L]  G. N. Lewis & J. E. Meyer. *Proc. Natl. Acad. Sci.*, **13**, 623–5 (1927).
[27.R]  O. K. Rice & H. C. Ramsperger. *J. Amer. Chem. Soc.*, **49**, 1617–29 (1927).
[28.P]  M. Polanyi & E. Wigner. *Zeit. physikalische Chem.*, **139**, 439–52 (1928).
[30.R]  O. K. Rice. *Zeit. physikalische Chem.*, **B7**, 226–33 (1930).
[32.K]  L. S. Kassel. *Kinetics of Homogeneous Gas Reactions.* The Chemical Catalog Company, New York. 1932.
[33.R]  O. K. Rice, *J. Chem. Phys.*, **1**, 375–89 (1933).
[34.C]  T. S. Chambers & G. B. Kistiakowsky. *J. Amer. Chem. Soc.*, **56**, 399–405 (1934).
[35.F]  A. Farkas. *Orthohydrogen, Parahydrogen and Heavy Hydrogen.* Cambridge University Press. 1935.
[38.B]  S. H. Bauer. *J. Chem. Phys.*, **6**, 402–3 (1938).
[38.F]  Faraday Society Discussion, Manchester, 1937. *Trans. Faraday Soc.*, **34**, 24–9; 70–81 (1938).
[39.B]  S. H. Bauer. *J. Chem. Phys.*, **7**, 1097–102 (1939).
[41.G]  S. Glasstone, K. J. Laidler & H. Eyring. *The Theory of Rate Processes.* McGraw-Hill, New York. 1941.
[45.C]  E. S. Corner & R. N. Pease. *J. Amer. Chem. Soc.*, **67**, 2067–71 (1945).
[45.H]  G. Herzberg. *Infrared and Raman Spectra of Polyatomic Molecules.* van Nostrand, Princeton. 1945.

## 1950–1959

[52.C]  C. F. Curtiss & J. O. Hirschfelder. *Proc. Natl. Acad. Sci.*, **38**, 235–43 (1952).
[52.K]  F. Kern & W. D. Walters. *Proc. Natl. Acad. Sci.*, **38**, 937–42 (1952).
[52.P]  H. O. Pritchard, R. G. Sowden & A. F. Trotman-Dickenson. *J. Amer. Chem. Soc.*, **74**, 4472 (1952).
[53.G]  B. G. Gowenlock, J. C. Polanyi & E. Warhurst. *Proc. Roy. Soc. London*, **A218**, 269–89 (1953).
[53.P1]  K. S. Pitzer. *Quantum Chemistry.* Prentice-Hall, Englewood Cliffs, NJ. 1953.
[53.P2]  H. O. Pritchard, R. G. Sowden & A. F. Trotman-Dickenson. *Proc. Roy. Soc. London*, **A217**, 563–71 (1953).
[53.P3]  H. O. Pritchard, R. G. Sowden & A. F. Trotman-Dickenson. *Proc. Roy. Soc. London*, **A218**, 416–21 (1953).

[53.R]   K. E. Russell & J. P. Simons. *Proc. Roy. Soc. London*, **A217**, 271–9 (1953).
[54.H1]  W. Heitler. *Quantum Theory of Radiation*. Clarendon Press, Oxford. 1954.
[54.H2]  J. O. Hirschfelder, C. F. Curtiss & R. B. Byrd. *Molecular Theory of Gases and Liquids*. John Wiley, New York. 1954.
[54.S]   R. G. Sowden. Ph.D. Thesis, Manchester University, Manchester. 1954.
[55.C]   E. A. Coddington & N. Levinson. *Theory of Ordinary Differential Equations*. McGraw-Hill, New York. 1955, pp. 62–78.
[55.S]   N. B. Slater. *Proc. Leeds Philos. Lit. Soc., Sci. Sect.*, **6**, 259–67 (1955).
[56.C]   D. Clark & H. O. Pritchard. *J. Chem. Soc.*, 2136–40 (1956).
[56.G]   B. F. Gray & H. O. Pritchard. *J. Chem. Soc.*, 1002–4 (1956).
[56.P]   H. O. Pritchard. *J. Chem. Phys.*, **25**, 267–70 (1956).
[57.W]   R. E. Weston. *J. Chem. Phys.*, **26**, 975–83 (1957).
[58.B]   F. W. Birss. *Proc. Roy. Soc. London*, **A247**, 381–9 (1958).
[58.G]   B. F. Gray & H. O. Pritchard. *J. Mol. Spectroscopy*, **2**, 137–43 (1958).
[58.L]   J. Langrish & H. O. Pritchard. *J. Phys. Chem.*, **62**, 761–2 (1958).
[58.M]   E. W. Montroll & K. E. Shuler. *Adv. Chem. Phys.*, **1**, 361–99 (1958).
[58.R]   B. S. Rabinovitch, E. W. Schlag & K. B. Wiberg. *J. Chem. Phys.*, **28**, 504–5 (1958).
[58.S]   E. W. Schlag. Ph.D. Thesis, University of Washington, Seattle. 1958.
[59.B]   A. N. Bose & C. N. Hinshelwood. *Proc. Roy. Soc. London*, **A249**, 173–9 (1959).
[59.G]   E. K. Gill & K. J. Laidler. *Proc. Roy. Soc. London*, **A250**, 121–31 (1959).
[59.H]   K. F. Herzfeld & T. A. Litovitz. *Absorption and Dispersion of Ultrasonic Waves*. Academic Press, New York. 1959.
[59.S1]  K. E. Shuler. *Phys. Fluids*, **2**, 442–8 (1959).
[59.S2]  N. B. Slater. *Theory of Unimolecular Reactions*. Methuen, London. 1959.
[59.W]   C. Walling & G. Netzger. *J. Amer. Chem. Soc.*, **81**, 5365–9 (1959).

## 1960–1964

[60.C]   J. P. Chesick. *J. Amer. Chem. Soc.*, **82**, 3277–85 (1960).
[60.S]   E. W. Schlag & B. S. Rabinovitch. *J. Amer. Chem. Soc.*, **82**, 5996–6000 (1960).
[60.W]   D. J. Wilson. *J. Phys. Chem.*, **64**, 323–7 (1960).
[61.C]   T. L. Cottrell & J. C. McCoubrey. *Molecular Energy Transfer in Gases*. Butterworths, London. 1961.
[61.F1]  W. E. Falconer, T. F. Hunter & A. F. Trotman-Dickenson. *J. Chem. Soc.*, 609–11 (1961).
[61.F2]  U. Fano. *Phys. Review*, **124**, 1866–78 (1961).
[61.P]   H. O. Pritchard. *J. Phys. Chem.*, **65**, 504–10 (1961).
[61.R]   O. K. Rice. *J. Phys. Chem.*, **65**, 1972–76 (1961).
[62.C]   N. Chow & D. J. Wilson. *J. Phys. Chem.*, **66**, 342–5 (1962).
[62.D]   N. Davidson. *Statistical Mechanics*. McGraw-Hill, New York. 1962.
[62.S]   F. W. Schneider & B. S. Rabinovitch. *J. Amer. Chem. Soc.*, **84**, 4215–30 (1962).
[62.W]   G. M. Wieder & R. A. Marcus. *J. Chem. Phys.*, **37**, 1835–52 (1962).
[63.B]   J. N. Butler & R. B. Ogawa. *J. Amer. Chem. Soc.*, **85**, 3346–9 (1963).
[63.C]   R. W. Carr, Jr & W. D. Walters. *J. Phys. Chem.*, **67**, 1370–2 (1963).
[63.G]   J. P. Galvin. Ph.D. Thesis, Manchester University, Manchester. 1963.
[63.K1]  A. D. Kennedy. Ph.D. Thesis, Manchester University, Manchester. 1963.
[63.K2]  A. D. Kennedy & H. O. Pritchard. *J. Phys. Chem.*, **67**, 161–3 (1963).
[63.K3]  G. H. Kohlmaier & B. S. Rabinovitch. *J. Chem. Phys.*, **38**, 1692–708; 1709–14 (1963).

[63.R] B. S. Rabinovitch, R. F. Kubin & R. E. Harrington. *J. Chem. Phys.*, **38**, 405–17 (1963).

[63.S1] F. W. Schneider & B. S. Rabinovitch. *J. Amer. Chem. Soc.*, **85**, 2365–70 (1963).

[63.S2] K. E. Shuler, G. H. Weiss & K. Andersen. *J. Math. Phys.*, **3**, 550–6 (1963).

[63.V] R. W. Vreeland & D. F. Swinehart. *J. Amer. Chem. Soc.*, **85**, 3349–53 (1963).

[64.A] K. Andersen, I. Oppenheim, K. E. Shuler & G. H. Weiss. *J. Math. Phys.*, **5**, 522–36 (1964).

[64.C] F. Casas, J. A. Kerr & A. F. Trotman-Dickenson. *J. Chem. Soc.*, 3655–7 (1964).

[64.E] D. F. Eggers, N. W. Gregory, G. D. Halsey & B. S. Rabinovitch. *Physical Chemistry.* John Wiley, New York. 1964, pp. 154–5.

[64.H] *Handbook of Mathematical Functions.* AMS 55, National Bureau of Standards, Washington, DC. 1964, p. 923.

[64.L] A. Lifshitz, H. F. Carroll & S. H. Bauer. *J. Amer. Chem. Soc.*, **86**, 1488–91 (1964).

[64.P] G. O. Pritchard, M. Venugopalan & T. F. Graham. *J. Phys. Chem.*, **68**, 1786–92 (1964).

**1965**

[65.H] M. L. Halberstadt & J. P. Chesick. *J. Phys. Chem.*, **69**, 429–38 (1965).

[65.P] D. W. Placzek, B. S. Rabinovitch, G. Z. Whitten & E. Tschuikow-Roux. *J. Chem. Phys.*, **43**, 4071–80 (1965); erratum, ibid., **44**, 3646 (1965).

[65.R] B. S. Rabinovitch, P. W. Gilderson & F. W. Schneider. *J. Amer. Chem. Soc.*, **87**, 158–60 (1965).

[65.W1] J. H. Wilkinson. *The Algebraic Eigenvalue Problem.* Clarendon Press, Oxford. 1965.

[65.W2] K. L. Wray. *Tenth Symposium (International) on Combustion*, The Combustion Institute. Pittsburgh, Pa. 1965, pp. 523–36.

**1966**

[66.B1] C. A. Brau, J. D. Keck & G. F. Carrier. *Phys. Fluids*, **9**, 1885–95 (1966).

[66.B2] D. L. Bunker. *Theory of Elementary Gas Reaction Rates.* Pergamon, Oxford. 1966.

[66.F] F. J. Fletcher, B. S. Rabinovitch, K. W. Watkins & D. J. Locker. *J. Phys. Chem.*, **70**, 2823–33 (1966).

[66.M] F. H. Mies & M. Krauss. *J. Chem. Phys.*, **45**, 4455–68 (1966).

[66.O] H. A. Olchewski, J. Troe & H. Gg. Wagner. *Ber. Bunsenges. phys. Chem.*, **70**, 450–9 (1966).

[66.P] D. W. Placzek, B. S. Rabinovitch & F. H. Dorer. *J. Chem. Phys.*, **44**, 279–84 (1966).

[66.T1] D. C. Tardy & B. S. Rabinovitch. *J. Chem. Phys.*, **45**, 3720–30 (1966).

[66.T2] D. M. Tompkinson & H. O. Pritchard. *J. Phys. Chem.*, **70**, 1579–81 (1966).

**1967**

[67.B] S. W. Benson & G. N. Spokes. *J. Amer. Chem. Soc.*, **89**, 2525–32 (1967).

[67.F] W. Forst, Z. Prasil & P. St. Laurent. *J. Chem. Phys.*, **46**, 3736–40 (1967).

[67.H] *Handbook of Tables for Mathematics*, 3rd Edn. Chemical Rubber Company, Cleveland. 1967, p. 842.

[67.R] D. G. Rush & H. O. Pritchard. *Eleventh Symposium (International) on Combustion*, Pittsburgh, Pa. 1967, pp 13–22.

[67.S1] N. B. Slater. *Mol. Phys.*, **12**, 107–10 (1967).

[67.S2] M. Solc. *Chem. Phys. Letters*, **1**, 160–2 (1967).

[67.W] W. D. Walters & D. J. Wilson. *Chem. Abs.*, **67**, 43209a (1967).

## 1968

[68.B1]  M. Bixon & J. Jortner. *J. Chem. Phys.*, **48**, 715–26 (1968).
[68.B2]  M. Boudart. *Kinetics of Chemical Processes*, Prentice-Hall, Engelwood Cliffs, NJ. 1968.
[68.L]   Y. N. Lin & B. S. Rabinovitch. *J. Phys. Chem.*, **72**, 1726–32 (1968).
[68.M]   T. I. McLaren & R. M. Hobson. *Phys. Fluids*, **11**, 2162–72 (1968).
[68.S]   D. H. Shaw & H. O. Pritchard. *Can. J. Chem.*, **46**, 2721–4 (1968).
[68.T]   D. C. Tardy & B. S. Rabinovitch. *J. Chem. Phys.*, **48**, 1282–301 (1968).
[68.V]   G. W. Van Dine & R. Hoffmann. *J. Amer. Chem. Soc.*, **90**, 3227–32 (1968).

## 1969

[69.A]   J. Aspden, N. A. Khawaja, J. Reardon & D. J. Wilson. *J. Amer. Chem. Soc.*, **91**, 7580–2 (1969).
[69.B]   H. E. Bailey. *Phys. Fluids*, **12**, 2292–300 (1969).
[69.F]   W. Forst & Z. Prasil. *J. Chem. Phys.*, **51**, 3006–12 (1969).
[69.H]   H. J. G. Hayman. *Trans. Faraday Soc.*, **65**, 2918–29 (1969).
[69.L1]  V. A. LoDato, D. L. S. McElwain & H. O. Pritchard. *J. Amer. Chem. Soc.*, **91**, 7688–93 (1969).
[69.L2]  R. L. LeRoy. *J. Phys. Chem.*, **73**, 4338–44 (1969).
[69.M1]  K. M. Maloney & B. S. Rabinovitch. *J. Phys. Chem.*, **73**, 1652–66 (1969).
[69.M2]  D. L. S. McElwain & H. O. Pritchard. *J. Amer. Chem. Soc.*, **91**, 7693–702 (1969).
[69.M3]  M. Menzinger & R. Wolfgang. *Angew. Chem. Int. Ed. Engl.*, **8**, 438–44 (1969).
[69.M4]  E. Meyer, H. A. Olchewski, J. Troe & H. Gg. Wagner. *Twelfth Symposium (International) on Combustion*. Pittsburgh, Pa. 1969, pp. 345–55.
[69.T]   T. F. Thomas, P. J. Conn & D. F. Swinehart. *J. Amer. Chem. Soc.*, **91**, 7611–16 (1969).
[69.W]   W. S. Watt & A. L. Myerson. *J. Chem. Phys.*, **51**, 1638–43 (1969).

## 1970

[70.B]   S. W. Benson & H. E. O'Neal. *Kinetic Data on Gas-phase Unimolecular Reactions*. NSRDS-NBS 21, National Bureau of Standards, Washington, DC. 1970.
[70.C]   S. C. Chan, B. S. Rabinovitch, J. T. Bryant, L. D. Spicer, T. Fujimoto, Y. N. Lin & S. P. Pavlov. *J. Phys. Chem.*, **74**, 3160–76 (1970).
[70.G1]  W. M. Gelbart, S. A. Rice & K. F. Freed. *J. Chem. Phys.*, **52**, 5718–32 (1970).
[70.G2]  R. A. Grieger & C. A. Eckert. *J. Amer. Chem. Soc.*, **92**, 7149–53 (1970).
[70.H1]  S. J. Hammarling. *Latent Roots and Latent Vectors*. University of Toronto Press, Toronto. 1970, pp. 30–2.
[70.H2]  M. R. Hoare & Th. W. Ruijgrok. *J. Chem. Phys.*, **52**, 113–20 (1970).
[70.K]   G. Kohnstam. *Progress in Reaction Kinetics*, **5**, 335–408 (1970).
[70.L]   Y. N. Lin & B. S. Rabinovitch. *J. Phys. Chem.*, **74**, 3151–9 (1970).
[70.M]   D. L. S. McElwain & H. O. Pritchard. *J. Amer. Chem. Soc.*, **92**, 5027–33 (1970).
[70.Y]   C. K. Yip & H. O. Pritchard. *Can. J. Chem.*, **48**, 2942–4 (1970).

## 1971

[71.F]   W. Forst. *Chemical Reviews*, **71**, 339–56 (1971).
[71.K1]  E. Kamaratos & H. O. Pritchard. *Can. J. Chem.*, **49**, 2617–19 (1971).

[71.K2] K. D. King, D. M. Golden, G. N. Spokes & S. W. Benson. *Int. J. Chem. Kinetics*, **3**, 411–26 (1971).
[71.M] D. L. S. McElwain & H. O. Pritchard. *Thirteenth Symposium (International) on Combustion*, Pittsburgh, Pa. 1971, pp. 37–49.
[71.O] I. Oref, D. Schuetzle & B. S. Rabinovitch. *J. Chem. Phys.*, **54**, 575–8 (1971).
[71.R1] O. K. Rice. *J. Chem. Phys.*, **55**, 439–46 (1971).
[71.R2] J. D. Rynbrandt & B. S. Rabinovitch. *J. Phys. Chem.*, **75**, 2164–71 (1971).
[71.Y] C. K. Yip & H. O. Pritchard. *Can. J. Chem.*, **49**, 2290–6 (1971).

## 1972

[72.B1] H.-J. Bauer. *J. Chem. Phys.*, **57**, 3130–45 (1972).
[72.B2] R. K. Boyd. *Can. J. Chem.*, **50**, 3104–8 (1972).
[72.D1] P. K. Davies & I. Oppenheim. *J. Chem. Phys.*, **56**, 86–94 (1972).
[72.D2] M. J. S. Dewar & M. C. Kohn. *J. Amer. Chem. Soc.*, **94**, 2704–6 (1972).
[72.D3] E. W. Dijkstra. *Comm. ACM.*, **15**, 859–66 (1972).
[72.D4] M. L. Dutton, D. L. Bunker & H. H. Harris. *J. Phys. Chem.*, **76**, 2614–17 (1972).
[72.E] G. Emmanuel. *Int. J. Chem. Kinetics*, **4**, 591–637 (1972).
[72.F1] W. Forst. *In: Reaction Transition States*. Ed. J. E. Dubois, Gordon & Breach, London, 1972, pp. 75–85.
[72.F2] W. Forst. *J. Phys. Chem.*, **76**, 342–8 (1972).
[72.F3] H. M. Frey, R. G. Hopkins & I. C. Vinall. *J. Chem. Soc. Faraday I*, **68**, 1874–83 (1972).
[72.F4] T. Fujimoto, F. M. Wang & B. S. Rabinovitch. *Can. J. Chem.*, **50**, 3251–4 (1972).
[72.G1] J. H. Gibbs. *J. Chem. Phys.*, **57**, 4473–8 (1972).
[72.G2] R. G. Gilbert & I. G. Ross. *J. Chem. Phys.*, **57**, 2299–305 (1972).
[72.J] D. W. Johnson, O. A. Pipkin & C. M. Sliepcevich. *Ind. Eng. Chem.*, **11**, 244–8 (1972).
[72.N] B. Noble, H. Carmichael & C. L. Bumgardner. *J. Phys. Chem.*, **76**, 1680–4 (1972).
[72.P] J. C. Polanyi & K. B. Woodall. *J. Chem. Phys.*, **56**, 1563–72 (1972).
[72.R] P. J. Robinson & K. A. Holbrook. *Unimolecular Reactions*. John Wiley, London. 1972.
[72.S1] T. Shimanouchi. *Tables of Molecular Vibrational Frequencies. Consolidated Volume 1*, NSRDS-NBS 39, National Bureau of Standards, Washington, DC. 1972.
[72.S2] G. B. Skinner & B. S. Rabinovitch. *J. Phys. Chem.*, **76**, 2418–24 (1972).
[72.W] E. V. Waage & B. S. Rabinovitch. *J. Phys. Chem.*, **76**, 1695–9 (1972).

## 1973

[73.A] T. Ashton, D. L. S. McElwain & H. O. Pritchard. *Can. J. Chem.*, **51**, 237–59 (1973).
[73.B1] T. Beyer & D. F. Swinehart. *Comm. ACM*, **16**, 379 (1973).
[73.B2] D. L. Bunker & W. L. Hase. *J. Chem. Phys.*, **59**, 4621–32 (1973).
[73.D] E. A. Dorko, R. W. Crossley, U. W. Grimm, G. W. Mueller & K. Scheller. *J. Phys. Chem.*, **77**, 143–8 (1973).
[73.F] W. Forst. *Theory of Unimolecular Reactions*. Academic Press, New York. 1973.
[73.G] D. M. Golden, G. N. Spokes & S. W. Benson. *Angew. Chem. Int. Ed. Engl.*, **12**, 534–46 (1973).
[73.K] E. Kamaratos & H. O. Pritchard. *Can. J. Chem.*, **51**, 1923–32 (1973).
[73.P] M. J. Perona & D. M. Golden. *Int. J. Chem. Kinetics*, **5**, 55–65 (1973).

[73.S]  S. E. Stein & B. S. Rabinovitch. *J. Chem. Phys.*, **58**, 2438–45 (1973).
[73.T]  J. Troe. *Ber. Bunsenges. phys. Chem.*, **77**, 665–74 (1973).


## 1974

[74.A]  G. Allen & H. O. Pritchard. *Statistical Mechanics and Spectroscopy.* Butterworths, London. 1974, pp. 64–6.
[74.C]  M. Christianson, D. Price & R. Whitehead. *J. Phys. Chem.*, **78**, 2326–9 (1974).
[74.D]  J. E. Dove, W. S. Nip & H. Teitelbaum. *Fifteenth Symposium (International) on Combustion.* The Combustion Institute, Pittsburgh, Pa. 1974, pp 903–14.
[74.F]  W. Forst. *Fifteenth Symposium (International) on Combustion.* The Combustion Institute, Pittsburgh, Pa. 1974, p. 680.
[74.L]  R. D. Levine & R. B. Bernstein. *Molecular Reaction Dynamics.* Clarendon Press, Oxford. 1974.
[74.M]  J. B. Moffat & K. F. Tang. *Theor. Chim. Acta*, **32**, 171–82 (1974).
[74.T]  J. Troe. *Fifteenth Symposium (International) on Combustion.* The Combustion Institute, Pittsburgh, Pa. 1974, p. 680.
[74.W1]  H. Gg. Wagner & F. Zabel. *Ber. Bunsenges. phys. Chem.*, **78**, 705–12 (1974).
[74.W2]  F-M. Wang & B. S. Rabinovitch. *J. Phys. Chem.*, **78**, 863–7 (1974).
[74.Z]  G. Zarur & H. Rabitz. *J. Chem. Phys.*, **60**, 2057–78 (1974).


## 1975

[75.J]  G. Jenner. *Angew. Chem. Int. Ed. Engl.*, **14**, 137–43 (1975).
[75.L]  S. H. Luu, K. Glänzer & J. Troe. *Ber. Bunsenges. phys. Chem.*, **79**, 855–8 (1975).
[75.N]  S. Nordholm & S. A. Rice. *J. Chem. Phys.*, **62**, 157–68 (1975).
[75.P1]  H. O. Pritchard. *Specialist Periodical Reports, Reaction Kinetics*, **1**, 243–90 (1975).
[75.P2]  I. Procaccia, Y. Shimoni & R. D. Levine. *J. Chem. Phys.*, **63**, 3181–2 (1975).
[75.P3]  I. Procaccia & R. D. Levine. *J. Chem. Phys.*, **63**, 4261–79 (1975).
[75.R]  P. J. Robinson. *Specialist Periodical Reports, Reaction Kinetics*, **1**, 93–160 (1975).
[75.T]  J. Troe. *MPT Int. Rev. Series, Phys. Chem.*, Series 2, Ed. D. R. Herschbach, Butterworths, London. 1975. Vol. 9, pp. 1–24.
[75.W]  F-M. Wang, Ph.D. Thesis, University of Washington, Seattle. 1975.


## 1976

[76.A]  M. H. Alexander. *Chem. Phys. Letters*, **38**, 417–21 (1976).
[76.C]  J. L. Collister & H. O. Pritchard. *Can. J. Chem.*, **54**, 2380–4 (1976).
[76.E]  J. Ernst, K. Spindler & H. Gg. Wagner. *Ber. Bunsenges. phys. Chem.*, **80**, 645–50 (1976).
[76.J]  K. Jug. *Theor. Chim. Acta*, **42**, 303–10 (1976).
[76.L]  D. K. Lewis. *Can. J. Chem.*, **54**, 581–5 (1976).
[76.N]  S. Nordholm. *Chem. Phys.*, **15**, 59–75 (1976).
[76.P1]  A. P. Penner & W. Forst. *Chem. Phys.*, **13**, 51–64 (1976).
[76.P2]  H. O. Pritchard & N. I. Labib. *Can. J. Chem.*, **54**, 329–41 (1976).
[76.P3]  H. O. Pritchard. *Can. J. Chem.*, **54**, 2372–9 (1976).
[76.R]  P. J. Robinson, personal communication.

## 1977

[77.A]   S. D. Augustin & H. Rabitz. *J. Chem. Phys.*, **67**, 64–73 (1977).
[77.B1]  M. H. Baghal-Vayjooee, A. W. Yau & H. O. Pritchard. *Can. J. Chem.*, **55**, 1595–8 (1977).
[77.B2]  R. K. Boyd. *Can. J. Chem.*, **55**, 802–11 (1977).
[77.G]   K. Glänzer. *Chem. Phys.*, **22**, 367–73 (1977).
[77.K]   M. Kuriyan & H. O. Pritchard. *Can. J. Chem.*, **55**, 3420–4 (1977).
[77.L1]  J. D. Lambert. *Vibrational and Rotational Relaxation in Gases.* Clarendon Press, Oxford. 1977.
[77.L2]  D. K. Lewis. *J. Phys. Chem.*, **81**, 1887–8 (1977).
[77.N]   S. Nordholm, B. C. Freasier & D. L. Jolly. *Chem. Phys.*, **25**, 433–49 (1977).
[77.O]   I. Oppenheim, K. E. Shuler & G. H. Weiss. *Stochastic Processes in Chemical Physics: The Master Equation.* MIT Press, Cambridge, Mass. 1977.
[77.P]   H. O. Pritchard. *Can. J. Chem.*, **55**, 284–92 (1977).
[77.Q]   M. Quack & J. Troe. *Specialist Periodical Reports, Gas Kinetics and Energy Transfer*, **2**, 175–238 (1977).
[77.R1]  K. V. Reddy & M. J. Berry. *Chem. Phys. Letters*, **52**, 111–16 (1977).
[77.R2]  O. K. Rice, personal communication.
[77.T1]  D. C. Tardy & B. S. Rabinovitch. *Chemical Reviews*, **77**, 369–408 (1977).
[77.T2]  J. Troe. *J. Chem. Phys.*, **66**, 4745–57; 4758–75 (1977).
[77.V]   L. Volk, W. Richardson, K. H. Lau, M. Hall & S. H. Lin. *J. Chem. Education*, **54**, 95–7 (1977).
[77.Y1]  A. W. Yau & H. O. Pritchard. *Can. J. Chem.*, **55**, 737–42 (1977).
[77.Y2]  A. W. Yau & H. O. Pritchard. *Can. J. Chem.*, **55**, 992–5 (1977).
[77.Y3]  A. W. Yau & H. O. Pritchard. *Can. J. Chem.*, **55**, 1588–91 (1977).

## 1978

[78.B1]  S. H. Bauer. *Chemical Reviews*, **78**, 147–84 (1978).
[78.B2]  R. C. Bhattacharjee & W. Forst. *Chem. Phys.*, **30**, 217–41 (1978).
[78.D]   J. E. Dove & J. Troe. *Chem. Phys.*, **35**, 1–21 (1978).
[78.F1]  M. C. Flowers. *Can. J. Chem.*, **56**, 29–31 (1978).
[78.F2]  K. F. Freed. *Accounts Chem. Research*, **11**, 74–80 (1978).
[78.H]   H. Hippler, K. Luther, J. Troe & R. Walsh. *J. Chem. Phys.*, **68**, 323–5 (1978).
[78.K]   I. E. Klein & B. S. Rabinovitch. *J. Phys. Chem.*, **82**, 243–5 (1978).
[78.N]   S. Nordholm. *Chem. Phys.*, **29**, 55–60 (1978).
[78.T1]  S. M. Tarr & H. Rabitz. *J. Chem. Phys.*, **68**, 642–6; 647–51 (1978).
[78.T2]  J. Troe. *Ann. Rev. Phys. Chem.*, **29**, 223–50 (1978).
[78.T3]  D. G. Truhlar. *J. Chem. Education*, **55**, 309–11 (1978).
[78.T4]  W. Tsang. *Int. J. Chem. Kinetics*, **10**, 599–617 (1978).
[78.T5]  W. Tsang. *Int. J. Chem. Kinetics*, **10**, 821–37 (1978).
[78.Y1]  A. W. Yau. Ph.D. Thesis, York University, Toronto, 1978.
[78.Y2]  A. W. Yau & H. O. Pritchard. *Can. J. Chem.*, **56**, 1389–414 (1978); erratum, ibid., **58**, 626 (1980), also Appendix 1.
[78.Y3]  A. W. Yau & H. O. Pritchard. *Proc. Roy. Soc. London*, **A362**, 113–27 (1978).
[78.Y4]  A. W. Yau & H. O. Pritchard. *Chem. Phys. Letters*, **60**, 140–4 (1978).

## 1979

[79.A1]  D. C. Astholz, J. Troe & W. Wieters. *J. Chem. Phys.*, **70**, 5107–16 (1979).
[79.A2]  S. D. Augustin & H. Rabitz. *J. Chem. Phys.*, **70**, 1286–98 (1979).

[79.B1] E. J. Beiting, G. F. Hildebrandt, F. G. Kellert, G. W. Foltz, K. A. Smith, F. B. Dunning & R. F. Stebbings. *J. Chem. Phys.*, **70**, 3551–2 (1979).
[79.B2] R. G. Bray & M. J. Berry. *J. Chem. Phys.*, **71**, 4909–22 (1979).
[79.B3] M. Buback & H. Lendle. *Z. Naturforsch.*, **34a**, 1482–8 (1979).
[79.C] R. T. Conlin & H. M. Frey. *J. Chem. Soc. Faraday I*, **75**, 2556–61 (1979).
[79.D] A. Dalgarno & W. G. Roberge. *Astrophys. J.*, **233**, L25–L27 (1979).
[79.F] W. Forst. *J. Phys. Chem.*, **83**, 100–8 (1979).
[79.G] J. Gelfand, W. Hermina & W. H. Smith. *Chem. Phys. Letters*, **65**, 201–5 (1979).
[79.H] H. Hippler, K. Luther & J. Troe. *Faraday Society Discussions*, **67**, 173–9 (1979).
[79.K1] D. F. Kelley, B. D. Barton, L. Zalotai & B. S. Rabinovitch. *J. Chem. Phys.*, **71**, 538–9 (1979).
[79.K2] J. H. Kiefer & J. C. Hajduk. *Chem. Phys.*, **38**, 329–40 (1979).
[79.K3] A. Kuppermann. *J. Phys. Chem.*, **83**, 171–87 (1979).
[79.M1] H. Mäder, H. Bornsdorf & H. Andersen. *Z. Naturforsch.*, **34a**, 850–7 (1979).
[79.M2] T. Minato, S. Yamabe, H. Fujimoto & K. Fukui. *Bull. Chem. Soc. Japan*, **51**, 1–10 (1979).
[79.O] I. Oref & B. S. Rabinovitch. *Accounts Chem. Research*, **12**, 166–75 (1979).
[79.P1] H. O. Pritchard. *J. Phys. Chem.*, **83**, 207 (1979).
[79.P2] H. O. Pritchard, N. I. Labib & A. Lakshmi. *Can. J. Chem.*, **57**, 1115–21 (1979).
[79.P3] H. O. Pritchard & A. Lakshmi. *Can. J. Chem.*, **57**, 2793–7 (1979).
[79.R1] K. V. Reddy & M. J. Berry. *Faraday Society Discussions*, **67**, 188–203 (1979).
[79.R2] K. V. Reddy & M. J. Berry. *Chem. Phys. Letters*, **66**, 223–9 (1979).
[79.R3] C. Rorres & H. Anton. *Applications of Linear Algebra*, 2nd Edn. John Wiley, New York. 1979. Chapter 7.
[79.T1] A. Talbot. *J. Inst. Maths. Applics.*, **23**, 97–120 (1979).
[79.T2] J. Troe. *J. Phys. Chem.*, **83**, 114–26 (1979).
[79.T3] J. Troe. *J. Phys. Chem.*, **83**, 149 (1979).
[79.Y1] A. W. Yau & H. O. Pritchard. *J. Phys. Chem.*, **83**, 134–49 (1979); erratum, ibid., p. 896.
[79.Y2] A. W. Yau & H. O. Pritchard. *Can. J. Chem.*, **57**, 1723–30 (1979).
[79.Y3] A. W. Yau & H. O. Pritchard. *Can. J. Chem.*, **57**, 1731–42 (1979).
[79.Y4] A. W. Yau & H. O. Pritchard. *Can. J. Chem.*, **57**, 2458–63 (1979).

### 1980

[80.C] W. E. Cooke & T. F. Gallagher. *Phys. Review*, **21A**, 588–93 (1980).
[80.D] J. E. Dove & A. Y. Grant, personal communication.
[80.E] G. E. Ewing. *J. Chem. Phys.*, **72**, 2096–107 (1980).
[80.F1] W. Forst & A. P. Penner. *J. Chem. Phys.*, **72**, 1435–51 (1980).
[80.F2] K. F. Freed & A. Nitzan. *J. Chem. Phys.*, **73**, 4765–78 (1980).
[80.G1] J. Gawlowski, T. Gierczak & T. Niedzielski. *J. Photochem.*, **13**, 335–46 (1980).
[80.G2] R. G. Gilbert & K. D. King. *Chem. Phys.*, **49**, 367–75 (1980).
[80.J1] R. L. Jaffe & H. O. Pritchard, to be published.
[80.J2] Th. Just & J. Troe. *J. Phys. Chem.*, **84**, 3068–72 (1980).
[80.L1] A. J. Lorquet, J. C. Lorquet & W. Forst. *Chem. Phys.*, **51**, 253–60 (1980).
[80.L2] W. J. LeNoble. *Review Phys. Chem. Japan*, **50**, 207–16 (1980).
[80.M] F. H. Mies. *Mol. Phys.*, **41**, 973–86 (1980).
[80.P1] H. O. Pritchard, G. M. Diker & A. W. Yau. *Can. J. Chem.*, **58**, 1516–26 (1980).
[80.P2] H. O. Pritchard. *Can. J. Chem.*, **58**, 2236–45 (1980).
[80.S] S. R. Singh & H. O. Pritchard. *Chem. Phys. Letters*, **73**, 191–3 (1980); erratum, Appendix 1.
[80.T] J. Troe. *Ber. Bunsenges. phys. Chem.*, **84**, 829–34 (1980).
[80.W] R. J. Wolf & W. L. Hase. *J. Chem. Phys.*, **73**, 3779–90 (1980).

**1981**

[81.C1]  P. Cadman & H. L. Owens. *J. Chem. Soc. Faraday I*, **77**, 3087–105 (1981).
[81.C2]  B. D. Cannon & F. F. Crim. *J. Chem. Phys.*, **75**, 1752–61 (1981).
[81.F]   W. Forst & S. Turrell. *Int. J. Chem. Kinetics*, **13**, 283–93 (1981).
[81.G]   J. Gawlowski & J. Niedzielski. *Int. J. Chem. Kinetics*, **13**, 1071–83 (1981).
[81.H]   H. Hippler, J. Troe & H. J. Wendelken. *Chem. Phys. Letters*, **84**, 257–9 (1981).
[81.I]   N. S. Isaacs. *Liquid Phase High Pressure Chemistry*. John Wiley, London. 1981.
[81.J]   J. Jortner & R. D. Levine. *Adv. Chem. Phys.*, **37**, 1–114 (1981).
[81.K1]  N. G. van Kampen. *Stochastic Processes in Physics and Chemistry*. North Holland, Amsterdam. 1981.
[81.K2]  K. D. King, T. T. Nguyen & R. G. Gilbert. *Chem. Phys.*, **61**, 221–34 (1981).
[81.L]   C. Lifshitz & E. Tzidony. *Int. J. Mass. Spect. Ion Phys.*, **39**, 181–95 (1981).
[81.M1]  M. A. Mohammadi & B. R. Henry. *Proc. Natl. Acad. Sci.*, **78**, 686–8 (1981).
[81.M2]  A. Moise & H. O. Pritchard. *Can. J. Chem.*, **59**, 1277–83 (1981).
[81.N]   S. Nordholm & H. W. Schranz. *Chem. Phys.*, **62**, 459–67 (1981).
[81.P1]  H. O. Pritchard. *Chem. Phys. Letters*, **78**, 618–20 (1981).
[81.P2]  H. O. Pritchard & S. R. Vatsya. *Can. J. Chem.*, **59**, 2575–86 (1981).
[81.R]   G. M. Rosenblatt. *Accounts Chem. Research*, **14**, 42–8 (1981).
[81.S1]  M. L. Sage & J. Jortner. *Adv. Chem. Phys.*, **37**, 293–322 (1981).
[81.S2]  M. S. Seshadri & G. Fritzsch. *J. Theoretical Biology*, **93**, 197–205 (1981).
[81.S3]  P. R. Stannard & W. M. Gelbart. *J. Phys. Chem.*, **85**, 3592–9 (1981).
[81.T]   H. Teitelbaum. *13th International Symposium on Shock Tubes and Waves*, Niagara Falls, N.Y. 1981, pp 560–9.
[81.V1]  S. R. Vatsya & H. O. Pritchard. *Proc. Roy. Soc. London*, **A375**, 409–24 (1981); erratum, Appendix 1.
[81.V2]  S. R. Vatsya & H. O. Pritchard. *Can. J. Chem.*, **59**, 772–8 (1981).
[81.V3]  S. R. Vatsya & H. O. Pritchard. *Chem. Phys.*, **63**, 383–90 (1981).
[81.V4]  A. A. Viggiano, J. A. Davidson, F. C. Fehsenfeld & E. E. Ferguson. *J. Chem. Phys.*, **74**, 6113–25 (1981).


**1982**

[82.A]   V. T. Amorebieta & A. J. Colussi. *J. Phys. Chem.*, **86**, 3058–60 (1982).
[82.B]   T. Baer & R. Kury. *Chem. Phys. Letters*, **92**, 659–62 (1982).
[82.C]   M. D. Clarkson & H. O. Pritchard, to be published.
[82.E]   D. van den Ende, S. Stolte, J. B. Cross, G. H. Kwei & J. J. Valentini. *J. Chem. Phys.*, **77**, 2206–8 (1982).
[82.F1]  W. Forst. *J. Phys. Chem.*, **86**, 1771–5; 1776–81 (1982).
[82.F2]  H. Furue & P. D. Pacey. *Can. J. Chem.*, **60**, 916–20 (1982).
[82.G]   R. G. Gilbert. *Int. J. Chem. Kinetics*, **14**, 447–50 (1982).
[82.H]   J. W. Hepburn, F. J. Northrup, G. L. Ogram, J. C. Polanyi & J. M. Williamson. *Chem. Phys. Letters*, **85**, 127–30 (1982).
[82.N]   W. B. Nilsson & G. O. Pritchard. *Int. J. Chem. Kinetics*, **14**, 299–323 (1982).
[82.P1]  C. S. Parmenter. *J. Phys. Chem.*, **86**, 1735–50 (1982).
[82.P2]  I. Powis. *Chem. Phys.*, **68**, 251–4 (1982).
[82.P3]  H. O. Pritchard & S. R. Vatsya. *J. Chem. Phys.*, **76**, 1024–32 (1982).
[82.P4]  H. O. Pritchard & S. R. Vatsya. *Chem. Phys.*, **72**, 447–50 (1982).
[82.P5]  H. O. Pritchard & S. R. Vatsya. *J. Comp. Phys.*, **49**, 173–8 (1983).
[82.R1]  K. V. Reddy, D. F. Heller & M. J. Berry. *J. Chem. Phys.*, **76**, 2814–37 (1982).
[82.R2]  T. R. Rizzo & F. F. Crim. *J. Chem. Phys.*, **76**, 2754–6 (1982).

[82.R3]  P. Rogers, D. C. Montague, J. P. Frank, S. C. Tyler & F. S. Rowland. *Chem. Phys. Letters*, **89**, 9–12 (1982).

[82.S1]  H. W. Schranz, S. Nordholm & N. D. Hamer. *Int. J. Chem. Kinetics*, **14**, 543–64 (1982).

[82.S2]  R. E. Smalley. *J. Phys. Chem.*, **86**, 3504–12 (1982).

[82.S3]  N. S. Snider. *J. Chem. Phys.*, **77**, 789–97 (1982).

[82.S4]  A. J. Stace & A. K. Shukla. *Chem. Phys. Letters*, **85**, 157–60 (1982).

[82.T]  A. B. Trenwith & B. S. Rabinovitch. *J. Phys. Chem.*, **86**, 3447–53 (1982).

[82.V1]  S. R. Vatsya. *J. Phys. A.*, **16**, 201–7 (1983).

[82.V2]  S. R. Vatsya & H. O. Pritchard. *J. Chem. Phys.*, **76**, 5171 (1982).

[82.V3]  S. R. Vatsya & H. O. Pritchard. *J. Chem. Phys.*, **78**, 1624–5 (1983).

[82.Y1]  W. Yuan, R. Tosa, K-J. Chao & B. S. Rabinovitch. *Chem. Phys. Letters*, **85**, 27–31 (1982).

[82.Y2]  W. Yuan, B. S. Rabinovitch & R. Tosa. *J. Phys. Chem.*, **86**, 2796–9 (1982).

# AUTHOR INDEX

# SUBJECT INDEX

activation energy, 7
  inequivalence with threshold energy, 38, 60, 77
  Tolman interpretation, 9, 58, 157
  variation with pressure, 9, 59–61
Arrhenius frequency factor, 7, 74
Arrhenius rate law, 7, 36
  *see also* non-Arrhenius
averaging, *see* grains, rotation

Born–Oppenheimer states, 81
bottleneck properties, 104, 121
  and strict Lindemann behaviour, 106, 111, 122, 157

cascade processes in relaxation, 77–8, 123
chemical activation, 99, 123
collision efficiency, *see* internal relaxation rate
collisionless regimes, 82, 100

decay rate constant, *see* specific rate function
  relation to vibration, 71
density of states, 35, 134
  steepest descent method, 136
  typical values, 34, 80, 82
detailed balancing, 1, 14, 41
disproportionation of radicals, 124

energy distribution in products, 66, 99
experimental inconsistencies, 112, 113

fall-off, 2, 28–9
fall-off position, 3, 50, 111
  effect of molecular complexity, 4, 5, 55, 111
  effect of temperature, 9, 57–9
  *see also* half-pressure
fall-off shape, 5, 50, 101, 110, 156
  effect of molecular complexity, 54
  effect of temperature, 9, 57, 156
  Lindemann, 5, 55, 101

fall-off shape, *cont.*
  sum of Lindemann forms, 46, 52, 62, 157
  strong collision, 41–58
  unusual, 87, 91, 97, 111, 157
  *see also* weak collision

Givens, see Householder
grains
  averaging over, 31, 68, 100, 137
  randomisation within, 83

half-pressure, 4, 51, 57, 115, 156
Householder transformation, 103

incubation times, 116
internal relaxation rate, 49–50, 121
  comparison with observed relaxation rates, 112
  dependence on assumed model, 105, 113–14
  independence of temperature, 59–61, 112, 114, 156
  *see also* surface activation
intramolecular relaxation, *see* randomisation
isomerisation – description of, xiii, 71, 93
isotope effects, xi, 5, 6, 87–90, 115

Kassel shape parameter $s$, 5, 56, 157

Laguerre quadrature, 49, 157
Lambert–Salter plot, 112
Laplace – inverse transform, 34–40, 47–8, 70, 73, 134
  *see also* partition function, specific rate function
Lennard–Jones interactions, 81–2
Lindemann mechanism, 1
  extended Lindemann expression, 56, 157
  *see also* fall-off shape, sum of Lindemann forms
Lindemann fall-off as a limiting shape, 108
  *see also* strict Lindemann behaviour
local modes of vibration, 66, 76

173